# Rekindling Life

Baptiste Morizot

# Rekindling Life

## A Common Front

Translated by Catherine Porter

polity

Originally published in French as *Raviver les braises du vivant.
Un front commun* © Actes Sud / Wildproject, 2020

This English edition © Polity Press, 2022

Polity Press
65 Bridge Street
Cambridge CB2 1UR, UK

Polity Press
101 Station Landing
Suite 300
Medford, MA 02155, USA

All rights reserved. Except for the quotation of short passages for the purpose of criticism and review, no part of this publication may be reproduced, stored in a retrieval system, or transmitted, in any form or by any means, electronic, mechanical, photocopying, recording, or otherwise, without the prior permission of the publisher.

ISBN-13: 978-1-5095-4927-6
ISBN-13: 978-1-5095-4928-3 (paperback)

A catalogue record for this book is available from the British Library.

Library of Congress Control Number: 2021951459

Typeset in 11 on 14 pt Fournier MT by
Cheshire Typesetting Ltd, Cuddington, Cheshire
Printed and bound in the UK by CPI Group (UK) Ltd, Croydon

The publisher has used its best endeavors to ensure that the URLs for external websites referred to in this book are correct and active at the time of going to press. However, the publisher has no responsibility for the websites and can make no guarantee that a site will remain live or that the content is or will remain appropriate.

Every effort has been made to trace all copyright holders, but if any have been overlooked the publisher will be pleased to include any necessary credits in any subsequent reprint or edition.

For further information on Polity, visit our website: politybooks.com

# Contents

*Acknowledgments*   vi

1 Give Us a Lever and a Fulcrum   1

2 Anatomy of a Lever, a Case Study: Hearths of Free Evolution   4

3 The Embers of Life   30

4 Realigning Alliances   51

5 Making Maps Differently: Dealing with Disagreements   133

6 Conclusion: The Living World Defends Itself   156

*Notes*   178
*Works Cited*   212

# Acknowledgments

For their invaluable rereadings, I would like to thank Frédérique Aït-Touati, Sarah Vanuxem, Sébastien Dutreuil, Sébastien Blache, Aurélien Gros, Bruno Latour, Jade Lindgaard, Charles Stépanoff, Christophe Bonneuil, Pierre Charbonnier, and Frédéric Ducarme.

Thanks to my editors: Anne-Sylvie Bameule for her confidence and benevolence; Aïté Bresson, Stéphane Durand, and Baptiste Lanaspeze for their remarkable editorial work; Actes Sud and Wildproject, two publishing houses I cherish that formed an alliance for the occasion.

Thanks to Alain Damasio for conspiring tirelessly on my formulations, and for bringing his friendship and his vitality as vectors to the project.

Finally, thanks – though the word is wholly inadequate to express what this book owes – once again, to Estelle Zhong Mengual, for her architectural genius, her powers of composition, her intellectual robustness. I hope she feels my gratitude on a daily basis for the moments we spend together on my projects, on hers, on ours – in our efforts to compose a reading experience that resembles a river, stringing together the pearls of ideas, arguments, and affects so that everything flows together, sometimes in torrents, sometimes meandering, sometimes jolting over the rapids, toward the sea that is life, out there.

I

# Give Us a Lever and a Fulcrum

On Monday, May 6, 2019, scientists working under the auspices of the Intergovernmental Science-Policy Platform on Biodiversity and Ecosystem Services (IPBES) submitted their report on the state of biodiversity: "Nature and its vital contributions to people, which together embody biodiversity and ecosystem functions and services, are deteriorating worldwide."[1] Action on the part of nation-states is quite obviously not commensurate with the stakes. Contemporary societies urgently need, through democratic means, to empower political structures capable of addressing this problem. At the same time, less institutionalized political forms, countless more local initiatives emerging from civil society, need to be deployed. Our collective intelligence has to take up the fight, in configurations yet to be invented, tested, profiled, propelled. A thousand initiatives are under construction right now, with little fanfare. Rebellions against extinction. Transformations in the use of territories. Cultural battles, too, over the meaning of words, the formulation of problems, the nature of our modern legacy, the prioritization of stakes.

Those of us who are aware of the crisis are growing in numbers. There is energy and intelligence to spare. We want to stop wasting time – we no longer have time to waste – on quibbles, purist posturing, fuzzy compromises, revolutionary romanticism: there are things to think and things to do (and often in that order, because nothing is more practical than a good theory).

## Rekindling Life

But a feeling of powerlessness prevails. The problem lies in the process of transmission between our hands and the world. We need ideas endowed with hands, and good ideas for the hands available.

The challenge comes down to inventing *levers*. A lever is an elegant arrangement; it is probably the first mechanism ever invented, the oldest of all. It is thought to have been developed several million years ago by our primate ancestors, who used their animal genius to launch into manual technologies (though other animals undoubtedly invented similar devices). The function of a lever is to make commensurate two things that are a priori incommensurable: on one side, a hand; on the other, a disproportionately large rock. By sliding a solid branch under the rock and wedging the rock against a solid support, a fulcrum, the animal intelligence that we have inherited can "move the earth," as Archimedes put it.[2]

We need *Archimedes' levers for large-scale ecological operations*, tools commensurate with the situation. We need them at the local level; we need them to be multiple, sharable, and effective. The lever is the only apparatus that can bring hands and the world into communication, that can make commensurable a set of actors (you, me, paltry little bit players) and the great adventure of life on earth, life that has been going on for billions of years – this biotic adventure that has made us all. For the action of ecological and evolutionary dynamics has chiseled us in every detail, with our opposable thumbs, the powers of love and curiosity that extend beyond our species (we see these powers even among certain cetaceans), our elegant and ambiguous primate brains, our political capacities for mobilization. All these powers are fundamental legacies of our evolution. Turnabout is fair play: it is time for us to mobilize our living powers to protect the adventure of life that has bestowed them on us.

## Give Us a Lever and a Fulcrum

I use the term "lever of ecological action" to name an arrangement that establishes commensurability between ourselves and the adventure of life all around us. A lever of ecological action must be efficient, accessible, uncompromising, rooted, effective in the short run and powerful *in the long run*; to achieve this latter goal, it must weave itself into the powers of resilience of life itself, those of the surrounding ecosystems. It must take on a real problem, a particular problem that can be overcome by a solution that is local but also scalable, applicable to a project that benefits global society (the lever and its "world"). Agroecologies that are plugged into short farm-to-table circuits are arrangements of this sort. In certain contexts, ZADs (Zones to Be Defended, an official designation in France) are also examples. There are thousands of levers to be invented. They are already sprouting up everywhere.

Here, I want to begin by exploring one local lever of this type, centered on the defense of forests. From there, we shall be able to move up a level in generality. For the philosophical and political analysis of the conflicts addressed in this case study will draw out a thread for further inquiry, an Ariadne's thread that we shall follow to try to get out of a labyrinth of dualisms, either/or dichotomies – nature/humanity, exploit/sanctuarize, wild/domestic – all of which create useless conflicts and keep us away from the real battlefronts.

We can now see the following question on the horizon: What becomes of "protecting nature," once we have understood that "nature" is a dualist construct that has contributed to the destruction of our milieus of life,[3] and that "protecting" implies a paternalistic understanding of our relations with the living world?

2

# Anatomy of a Lever, a Case Study: Hearths of Free Evolution

The project that I propose to investigate here seems to me to have several properties that characterize a lever for large-scale ecological action; it is an ideal example, as we shall see. It targets a specific problem. It is local, but it is powerful. It has to do precisely with the drama of species loss, the weakening of ecosystems induced by fragmentation of milieus, by overharvesting and overhunting. It responds to these challenges as best it can – for the time being, microscopically, but already effectively, at its level. Above all, it is real.

I am referring to a project calling for the radical protection of what are known as hearths of free evolution, protection ensured through the legal and economic tool of land acquisition. Initiatives of this sort in France were initially undertaken under the auspices of the Forêts Sauvages (Wild Forests) association, and they are now supported by the Association pour la Protection des Animaux Sauvages, or ASPAS (Association for the Protection of Wild Animals).[1] In the 2019 Vercors Vie Sauvage (Wildlife in the Vercors) initiative, the case on which we shall focus here, ASPAS acquired a forest of 500 hectares (about 1,235 acres) in the gorges of the Lyonne river valley.

For what purpose? To leave it alone. Restore it to the beech trees, silver firs, deer, squirrels, wolves, eagles, tits, lichens: to wild prairies and mature forests. Leave it in free evolution – that is, let the milieu develop according to its own laws, without exploiting, modifying, or managing it. Leave the dead trees standing so they can become

## Anatomy of a Lever, a Case Study

habitats for other living beings. Leave the fallen wood on the ground so it will melt into humus. Let all manner of living creatures come and go. Let evolution and ecological dynamics do their serene, stubborn work of resilience, invigoration, circulation of energy, creation of life forms. Cut short all "anthropic forcing."[2] These preserves are open to humans: anyone can go in, provided they respect the site.

The idea is diabolically simple. It does not look very revolutionary, but it harbors legal displacements, political subversions, and philosophical decisions that we shall explore in depth. It emerges at the confluence of three concepts (for the originality of an idea often lies in its being the meeting place and unique knotting point of other ideas): free evolution (as the style of management of a specific milieu), land acquisition by a nonprofit association (as a means of making protection permanent), and participatory financing (as a way of mobilizing the citizenry to share in joint ownership).

### From small lives to Great Life

The singularity of the project that I am seeking to track here lies in its relation to time. When one is on the territory of the recently acquired nature reserve Vercors Vie Sauvage, for example, one can appreciate the temporal scales of other living things. The beechnut that just fell at my feet contains four seeds; one of them could become a venerable beech tree if it sprouts tomorrow, if it is not cut down, if it is allowed to live its lives. It will be the wild forests of tomorrow, the ancient forests, the richest milieus, the most timeless. If we allow it the time, it will become a habitat tree sheltering a proliferation of fauna: a whole cosmopolitan world will dwell in this cosmos, with its differentiated floors, its multiform communications, its labyrinth of unknown lives, its interspecies conventions. In this forest, there are already a few beech trees that may well be a couple of centuries old. Under their branches, one feels what it means to build a world, a world for the other forms of life. From the sprouting of one fragile seed to the mastodon before our eyes, the life of this beech is like a very slow explosion that goes on and on for centuries. It is an expanding galaxy that welcomes and shelters all reigns, from squirrels to lichens. A very slow explosion that undertakes stunning formal quests in order to explore ways of dialoguing with the elements: air, water, earth. Experimenting with the world, feeling its way, from the tips of its branches and its roots, with its infinitely slow intelligence. Taking centuries to explore, by palpations of sky and rocks, the possibilities of being a tree. It is this sort of tree that can flourish and repopulate a preserve like Vercors Vie Sauvage. It is this sort of forest, this type of Great Life, that the sites of free evolution are meant to resuscitate. Nothing more, nothing less.

But doing so will take 300 years at the very least. The ecologists explain that biodiversity abounds in a tree after 100 or 150 years.[3] In Europe, a third of the biodiversity that a tree harbors depends on the later stages: this is when it really becomes a world for myriad other life forms. Exploited trees *never* reach such an advanced age; allowing them to do so is not economically viable, according to the criteria of contemporary forestry.

Our longevity as human individuals is trivial in comparison to that of a tree, a coral, an ancient forest, an ecosystem. Yet the Great Life of ecosystems, the green lungs of forests, carbon cycles, the evolution of species, is the condition enabling the small lives of individuals. What is at stake in a lever for ecological action is protection of that Great Life. But in order to protect something, we are compelled to see the world from the vantage point of what we want to protect. For we can only protect a forest by protecting its world, and we can only understand its world by grasping time and space from the perspective peculiar to this life form. By following its own way of fashioning its space-time. To truly protect something is to protect it from *its own point of view*. Indeed, its own point of view is what we must protect.

A defining feature of this Great Life is that it lives and breathes on the scale of centuries and millennia. We must therefore protect it in the same dimensions.

While our light bulbs are designed to last six months, our policies devised to last a few years, why not imagine a politics for life forms that would be conceived on the scale of centuries?

That is the untimely ambition of sites of free evolution that have been established through land acquisition: to bring into being the ancient forests of tomorrow. The idea is to protect feral nature, the nature that regenerates spontaneously if we leave it to act on its own. But this nature has to be protected where people live – in the Drôme, in the Massif Central, in Brittany – in order to engage local populations; if we protect only prestigious and remote natural sites (isolated national parks, sublimely high mountains), we are implicitly justifying the abandonment of *all* other milieus.[4]

But how are we to *act*, *now*, *urgently*, on the scale of centuries – while lobbies are pressuring us to extract resources, to open new spaces for exploitation, to cut down every tree as soon as it is 60 years old, in the mad race to keep the markets going?

## *A politics of the living on the scale of centuries*

Here is where the genius of these projects comes in: the trick is to take advantage of French property law and subvert it. Divert it, inasmuch as it is one of the causes of the ecological crisis: land ownership is what guarantees to exploiters the right to pressure the milieu for their private interest, sometimes to the detriment of the web of life. The idea here is to use property law to fight *against* such deviations. Article 544 of the French Civil Code consecrates ownership as the right of the owner to "enjoy and dispose of things in the most absolute manner." This right is, in part, what makes it possible to weaken and sometimes devastate spaces in the name of profitability. "Absolute" does not mean "sovereign" (the right to do anything and everything), because the right of ownership is framed by the formula "provided that [the owner] does not use it in a way prohibited by laws or regulations." An "absolute right" means, here, a "deterrent" right: in other words, a right opposable to all, a right that allows non-owners to be prohibited from using the property.

But if ownership gives the right to absolute exploitation of a milieu, limiting external control, this means that it also gives the right to absolute *protection*, without suffering outside pressure from lobbies. The idea is to take advantage of the possibilities offered by the property law and turn it against itself, against its world. To carry out an infiltration in plain sight.

Any attempt to create a national or regional nature reserve is exposed to countless demands by hunters, farmers, foresters, livestock breeders and herders, industrialists: all of them refuse to see public land put out of reach of their multiform exploitations (extracting, pasturing, logging, harvesting, hunting, and so on). These negotiations among different users of a given territory are important and appropriate in most contexts. Émilie Hache has recently written an important book on the necessity of such negotiations and the shape they ought to take, as a democratic form of the relation between humans and their milieus.[5] A priori, and in general, we might suppose that these negotiations are the best option, especially to prevent the risk of local populations losing their lands in the name of protecting nature. But

*Anatomy of a Lever, a Case Study*

moving beyond general positions, we have to look closely at the contexts and situations to do justice in each case. For, in the French context we are focusing on, when it comes to protecting patches of forest or rivers, brandishing negotiation as a moral principle is in fact one more weapon exploiters can use to prevent even the simplest and most reasonable measures for protecting milieus: when the power relation is too unequal, defending negotiation amounts to defending the powers that be. (I shall show this further on by clarifying the logic of *unequal spatial scales*.) It means defending the parties with the greatest economic and political lobbying power.[6] And, as we all now know, this regularly occurs to the detriment of the common good – common to humans and the other living beings.[7]

Protectors of nature have thus been impotent witnesses to measures as contradictory as the return of hunting or pasturing in the very heart of some national parks or in the best-preserved zones of certain regional parks, which are already microscopic. They have similarly watched conservatories of natural spaces gradually introduce *active management* into sites that were once in free evolution. These spaces are being modified according to the logic of heritage sites, according to aesthetic criteria, security concerns, and/or with the aim of protecting certain targeted species. (This conservation model can be defended in certain contexts, provided that there are no claims to monopoly.)

In response, "ASPAS, not being satisfied with the policy governing protected areas, and the deviations from it that have become customary, has created a new status, which corresponds to status 1b, 'Zone of wild nature' of the UICN (International Union for Conservation of Nature) and has registered the name."[8] The status of Wildlife Reserve (*Réserve de vie sauvage*) applies to areas in free evolution. It is simply a matter of taking ownership. The challenge is to skirt a double ambush: the compromises induced by unequal power relations with exploiters on one side, the shifting policies in the management of protected species on the other. The Forêts Sauvages association has adopted analogous goals, using different terminology. It hardly matters – there may be multiple strategies, but the project is what counts.

Land acquisition is precisely what makes it possible to put a stop to endless compromising with lobbyists acting on the part of would-be exploiters. Private ownership in fact allows acquirers to avoid most of these tensions and negotiations: their enjoyment is "absolute," in the legal sense of "actionable in opposition to all non-owners."

The first key idea behind the project of creating sites of free evolution is thus to turn the legal invention of private property to the benefit of life forms other than our own: to the benefit of other living beings besides individual human proprietors. Because our law has been carved out by and for the possessors, property law is, paradoxically, a major weapon for protecting milieus: it suffices to turn the situation upside down. Ownership gives absolute enjoyment to the owner, but here the owners are not buying in order to secure personal enjoyment – they are buying to restore enjoyment to *other forms of life*.

No one will be able to cut down trees to sell them cheaply; no one will be able to sort out the timber, pry out the badgers, feed the deer corn so as to be able to shoot them on sight: the territory will be left to itself. Ecologist and forester Alain Persuy asks: "Are we ready, alongside forests that are exploited in multifunctional ways, to leave some of them in peace?"[9]

In other words, to leave the forest in free evolution. In 2008, Jean-Claude Génot proposed "free evolution" as a management style for natural spaces in France.[10] In *La nature malade de la gestion* (Nature

## Anatomy of a Lever, a Case Study

Suffering from Management), Génot offered a critique of the shift in management practices that began to take hold in the 1970s, a shift toward active intervention in and modification of wild milieus set aside for protection; this was a landscapist, garden-oriented approach to conservation, incapable of accepting the ascetic posture of *doing nothing*. While modification may make sense in some conservation initiatives on the local level, Génot was criticizing its institution as the dominant and generalized logic for protecting milieus.

In contrast, a territory in free evolution is a space-time where diversity is allowed to settle in spontaneously: that of individuals (in terms of age or conformation), species (many exploited forests harbor a single targeted species), forms (vines, underbrush, strata), and dynamics of landscape creation and successions (a damp zone has a tendency to be colonized by willows over time, and then to become a forest; an uprooted tree leads to an explosion of sun-loving species).

It is not a question of following the American tradition of preserving ecosystems as they would have been before the arrival of humans (while forgetting the role of Amerindians, moreover, in shaping American landscapes), in a state of supposed patrimonial virginity. Free evolution, the inverse of the cult of wilderness as nature pristine and intact, accepts the human history of forests.[11] In Europe, forests are often interwoven with complex ancient human ways of exploiting them. The point is not to turn back toward a supposed purity, but to allow the spontaneous forces of the forest to take over again. This is what is called ferality: letting an ecosystem express its powers, use its capacity to regenerate itself after it has been transformed by humans.

Leave – that is, restore – wild life to itself. This is the second key idea in all its troubling beauty. A forest in free evolution does what life does. It struggles spontaneously against global warming, by limiting the greenhouse effect. It stocks carbon, and it does this all the better when its trees are ancient and venerable. It works to purify water and air, to form soils, to reduce erosion, to foster a rich, resilient biodiversity capable of absorbing the blows of the coming climate changes. It does not do this for us, but it does it all the same, and its gifts are priceless.

Why reason in terms of profits and losses if, here, everything is offered and impregnable?

## *In free evolution*

The idea of "leaving the forest to itself" triggers traumatic echoes in many people. We need to start by disrupting the associations. For the idea is taken indiscriminately today (by the "deciders" first and foremost) as implying rural desertification, loss of control over the territory, the erasure of human presence, an invasion of wildness – and everyone wants to fight against it, without quite knowing what is involved. Because the idea of a fragment of the world left to itself is terrifying.

But the real problem lies elsewhere, in the matter of spatial scale, which must never be forgotten. For it is not the "world" in general that would be left to itself, restored to itself, but only morsels of wild life in a French territory of which 99% has been exploited, transformed, hunted, anthropized. What these preserves are currently trying to protect from destructive human activities are tiny spots – confetti scattered over the map. The zones truly protected from exploitation, appropriation, and development in France oscillate between 0.02% and 1% of the French territory, depending on the criteria used for measuring. Humans can already manage, develop, sometimes cut down, dry up, and build, virtually *everywhere*. Is it really so unreasonable to imagine restoring a few parcels

## Anatomy of a Lever, a Case Study

of peace and quiet to the other life forms that populate the Earth alongside us?

One model of scientific ecology makes the disparities readily visible. In substance, this model consists in comparing the biomass of vertebrates (let us say animals in general, humans included) on the surface of the Earth, on two dates: 10,000 years ago, and today. Of the animal mass 10,000 years ago, 97% was constituted by wild fauna, with humans weighing in at about 3%. Today, domestic animals make up 85% of the biomass of all terrestrial vertebrates, and humans have moved up to 13%. Wild fauna, once 97% of the total, now constitute just 2%.[12] A massive overturning, a colossal confiscation of the biomass by domesticated animals, to the detriment of the other components of the ecosystems – wild fauna, in particular. In the process, humans have amputated 50% of the autotrophic biomass (plants, let us say).[13] These numbers do not require lengthy commentaries. We can let them settle in our minds, where they can work at turning us into living beings *among others*.

And yet the defenders of exploitation continue to stigmatize all solid efforts to protect milieus, and they continue to demand compromises, exploitation rights even within the protected confetti. Gilbert Cochet described the phenomenon one day when we were exploring the western portion of the Vercors Vie Sauvage reserve:

> It is as though, by sharing the wealth between the exploiters and nature, we were giving 99% to the former and 1% to the latter. But then the exploiters come in and say: "In your 1%, you have to compromise with us, you have to allow economic activities, otherwise it's unfair: we can't give everything to nature." But they already have virtually all the territory![14]

## Rekindling Life

Whatever the acolytes of growth may say about these proposals for forests in free evolution, there is no secret plan for covering the whole world with preserves in which exploitation would be forbidden. There is no all-powerful hidden eco-tyrannical conspiracy determined to forbid the world to humans: it is from the standpoint of pugnacious underlings that a minority is standing up for zones of free evolution. The objections to making sanctuaries of these plots are in fact ideological: they reverse the dominant and the dominated parties. The real power relation is the inverse: protecting these spaces and their fauna is a struggle like David's against Goliath.

The resistance is simply seeking to withhold a small percentage of land from exploitation, to the benefit of the living fabric that constitutes our giving environments. Is this such a radical quest, or simply a matter of good sense and a touch of decency? These days, it seems that humanism itself has changed sides.

The second fear associated with free evolution is fear of a "return of wildness": fear of losing control over the territory, giving up our role as "developers of the earth," being "submerged" by wildness. This fantasy is easily neutralized if we look at a phenomenon that is already omnipresent: all the landowners who are practicing free evolution "unawares." In fact, millions of acres of French forest land are the private property of people who are unaware of the fact, or pay no attention to it, having inherited the land in some vague way; these sites are already peacefully undergoing free evolution. Still, if any owner should decide to profit from them or modify them, free evolution and its powerful long-term effects would be wiped out. For the decline in economic and managerial activity related to forests left to themselves is only the current phantom of past exploitation: we need to think in terms of free evolution as a *positive* development, one that does not entail simply an act of abandonment but rather an affirmative act of protection over the long term that would make the forested lands in question sanctuaries immune to volatile cycles of multiform exploitation. The word "protection" is actually not the right one, here, as we shall see: it is rather a matter of invigoration, dynamic conservation centered on ecological and evolutionary potentials given the conditions that allow them to flourish.

Forests in free evolution do not imply the return of a wilderness that would submerge civilization; they are just forests developing as they do when they are forgotten, when people stop thinking that they have to be developed, valorized, modified so they can flourish. Nor is free evolution a straightforward letting-go: it is one diplomatic practice among other possible practices that could be adopted toward forests. It consists in recentering our viewpoint toward that of a forest, then taking seriously its own patterns of behavior, and finally seeking the best way to give the forest in all its richness the conditions under which it can express itself. This is a strong position, but one that is difficult to maintain in the recent culture of conservation, obsessed as it is by blind faith in the need to take charge, sometimes in defiance of good sense (for example, when managers of natural milieus are compelled by law to cut down all the dead trees, to protect the "safety" of people walking in the forests, thereby depriving all the fauna of innumerable habitats offered by the cavities in standing dead wood: a sort of protection *against* nature). Free evolution is a specific, positive, thought-out practice: it means doing nothing at all so that a forest is left free to regenerate.

The initiative to defend forests in free evolution is thus a subtle arrangement for navigation that tacks between problematic approaches, weighty inheritances, and abstract categories that are stigmatized as a whole. The project behind the establishment of hearths of free evolution in France entails three shifts away from the currently dominant trends: it avoids setting up a patrimonial "virgin" nature – a notion proper to the American tradition – by betting on the strengths of ferality and naturality; it uses land acquisition to neutralize unsustainable exploitation; and it fends off the impulse to intervene by adopting the principles of free evolution.

## *Deconstructing a slogan: "nature under glass"*

The widespread, almost automatic image summoned up by the terms "nature reserve" or "integral protection" is that of nature under glass. This particularly insidious expression warrants painstaking deconstruction.

Put under glass, put in a glass case: these dismissive expressions are brandished on a massive scale by the agrobusiness and pro-hunting lobbies, in order to devalorize, in the eyes of citizens and politicians, all attempts at robust protection of nature – attempts dreamed up by people concerned about portions of milieus that are in jeopardy, reaching a breaking point, according to scientific reports. In the French context, the expressions in question originally referred to the patrimonialization of natural sites as if they were set in glass cases, protected from the vast intentional upheavals of massive industrialization of the landscape; the formulas have since been instrumentalized by the adversaries of all robust forms of milieu protection, opponents of any form of limitation on the power of those who exploit our living spaces.

Such slogans serve to activate three fantasies. First, that of a fixed patrimonialization, which would petrify and isolate nature as if it were in a bell-jar, with an imagined rigidity, or "fixism."[15] This makes no sense as applied to the reserves in question, since the point of establishing such spaces is to return them to free evolution, so they can reconstitute their evolutionary potential and their ecological dynamics: it is, rather, a matter of leaving ecosystems in peace so they can *transform themselves* according to their own internal logics. And this is where they unfold in their full plenitude. As Gilbert Cochet says, "leaving forests in free evolution is like unshackling athletes' legs: they begin to run."[16]

The second fantasy activated by watchwords such as "nature under glass" is even more problematic: it is the fantasy of the theft of a common good – the idea that these protected spaces are *stolen* from rural populations, that the locals are robbed of what "belongs" to them, "their" forests, "their" mountains. In fact, in the first place, we are still talking about very small parcels relative to the territory as a whole. Next, these are not agricultural or pastoral lands, they are essentially forested slopes, wilderness zones. Finally, restoring these small plots to other living beings does not mean excluding humans from them: anyone can go there to be immersed in a rich life, to experience self-recognition by observing other forms of life, to celebrate life in all its forms. Reserves are accessible; anyone may enter. On a small signboard at the entrance to the Grand Barry preserve, acquired by ASPAS in 2012 and in free evolution since then, we can read, for example: "You are welcome in this space in free evolution; please respect its integrity."

Everyone can come in, then, leaving tools and weapons at the door (is that not how one enters a space of diplomacy?); the glass case is open. And, above all, all living beings within it can *go out*: a wildlife preserve constitutes a space of regeneration in which life takes back its rights and proceeds to irrigate all the surrounding territory with vitality. It is a hearth, and as such it is a *source*, an overflowing spring from which life trickles out toward its surroundings (unlike the notion of a trickle-down economy, there is in fact a theory of trickling that is accurate – that of scientific ecology).

So, in free evolution there is no question of putting "nature under glass." Indeed, the purpose is exactly the opposite: it is a matter of creating a green heart in a territory so that vigorous life can be diffused everywhere in the vicinity. Conservation ecology teaches us that life reconstructs its adaptive potentials, puts itself back on an ascending ecological trajectory, begins to flourish once again, as soon as it is given the space and time to rekindle its multiform fire. This vitality is truly destined to spread, for a hearth of free evolution begins with a single practical measure: removing the fences.

Humans have the right to come and go, and all nonhumans have the right to leave: pollen from the trees; grains in birds' crops; wild pollinators that are largely responsible for the success of our harvests; birds that are in decline elsewhere but here can build their nests in standing dead trees; dispersing river otters; deer and chamois calmed and strengthened by the sanctuary. From here, all life can spread out into the world that has been damaged by blind exploitation. Free evolution does not mean confiscating a space; it means invigorating all the wildlife of a territory through the maintenance of small kernels of life where wild beings can gain strength and then disseminate everywhere, in the form of flowers, insects, beavers, eagles, and birds of the fields whose populations are being destroyed by intensive agriculture. All these forms of life can go out and repopulate the surrounding exploited world, restoring to it a more complete, more resilient, richer biodiversity.

A hearth of free evolution thus provides a common benefit, shared, offered, and unobstructed. A *common good*, vital in this period of intensive destruction of biodiversity. Common, first of all, because it is open to all, and because it is an initiative that works to the benefit of all life forms across interspecies barriers (lichen, deer, pollinators, human inhabitants are interwoven without distinction in this common). It is a multispecies common good, constructed – an intriguing paradox – on the basis of private property law.

What is defended in such a hearth is a collective body, and this collective body is multispecies, it is the forest itself, its members endowed with feathers, fur, or leaves.

Wild forests protected by preserves are something like havens from which all the forces that destroy the web of the living are banished, so that the forests can reconstitute and reinvigorate themselves. There will be compromises to be made with the life forms that come into conflict with human practices: fierce negotiations, illusion-free diplomacy, will have to work toward cohabitation. Vigorous hospitality will be called for: offering welcome even while resisting. This is the self-evident oxymoron of life – it is the relationship invented millions of years ago in bodies (the immune system) and in ecosystems (the ecological interactions that, even if they are mutually beneficial, require knowing how to defend oneself against abuse by participating partners); this is the type of negotiation we need today.

### *At home, but first in the homes of so many others*

In a forest in free evolution, then, you can enter as a human mammal. You are welcome whenever you like, knowing that for once you are first and foremost in the homes of others, in their hearths, on their familiar paths. In a sense, you are also in your own home, but not as an owner: this time, rather, as a cohabitant of the earth.

Under these conditions, one may wonder why it is forbidden to pick wild strawberries and mushrooms or to gather dead wood in a wildlife preserve. There are two ways this restriction might be justified: out of principle, or out of precaution. It could be forbidden *on principle*, based on the idea of restoring these spaces as they existed before humans came along. But this principle is actually anomalous. Let us take the example of the Vercors Vie Sauvage forest: it emerged after the last glaciation, presumably between the Upper Paleolithic and the Mesolithic eras, which means that the development of the fir–beech forest *followed* the arrival of humans in the territory that came to be called the Vercors (*Homo sapiens* arrived in what is now France more than 40,000 years ago). In other words, these forests have always known human inhabitants, hunter-gatherers who gleaned plants, berries, and firewood, at first probably in populations of very low densities. Thus, to forbid all food-related activities on principle would be to create a fantasy space that has never existed. The fantasy is based on a dualist and misanthropic conception of the place of humans in ecosystems – as if they were of a different nature from nature, so that all human activity would be polluting and degrading. But this dualism is mistaken: as living beings among others, as mammals, we have always been dispersers of seeds, and pollinators. From the epoch when we walked barefoot through tall grass to the epoch of hiking boots, we have always carried along – at our ankles and in our hair in the era of bare feet, in the folds of leather moccasins in the Paleolithic era, and in our shoes and socks today – pollens destined to fertilize plants all along our paths. Zoochoric plants have always used us as blind vessels, oblivious to our role, in order to exercise their nomadic sexuality. By collecting fruit and medicinal herbs, we have also participated in the dispersal of seeds and plants. We are foxes like the others.

If we forbid gathering and gleaning in a forest, we demonize the whole part of humanity that connects it with ecosystems – in other words, its role in the trophic chain, and its historical ecological roles. Why, then, does the charter of the wildlife reserves in France (RVS) forbid taking anything from these territories? It is a matter of precaution. In principle, there is no reason to prevent a few people from gathering mushrooms and raspberries in an RVS (these people are mammals like any other). But the problem is simple: we do not know and cannot control the number of gatherers, nor would we want to establish limits, since the spaces are open – there might be 3 gatherers a year, or 3,000. But, as soon as we go beyond a certain threshold, these takings damage the forest dynamics by exceeding the load and regenerative capacities of the milieu. The problem is not the act itself, it is the number of repetitions. Out of *precaution*, then, we can decide, pragmatically, that nothing will be removed; this will allow us to see the forest unfold without risk. It is not a matter of depriving humans; we must not forget that everywhere around this space, in almost the totality of French territory, taking things from the milieu is not only possible but prevalent, and it exacts a price: its imprints are too massive, its effects often deleterious.

The decision to forbid the removal of anything from the forest thus results not from putting the forest under glass, in a fantasized encroachment on our individual freedoms, but from adopting local and circumstantial measures that seek to anticipate everything that might hinder the regeneration of the ecologic functionalities of the site.[17] To speak of a "forest under glass" is to use a rhetorical tool aimed first of all at arousing among those who hear it the panic of being deprived of a space – the panic of not being able, for once, to be the sovereign ruling over what exists.

The third and final dimension of the fantasy of putting nature under glass emerges here: the protectors of nature are accused of "wanting to forbid everything," often by exploiters who set themselves up as defenders of our freedoms, arousing in the hearer the indignation of a society jealous of its subjective rights. It is easy to respond factually to that accusation: here, in this little hearth of free evolution, you have the right to do *everything* – except to exploit, take away, kill, damage, and endanger the integrity of the space.

## Facing climate change

Hearths of free evolution are neither closed nor abandoned nor petrified. They are not dead spaces, taken out of the territory's common pot, that *lie idle* and fade away: they are living hearts that gleam and overflow every spring. They are not spaces that are *put* to work, but spaces in which "something" is at work; they are very active spaces, embracing variation, creation, ripening of forms, interweavings, colossal production of biomass through carbon storage, water purification, climate stabilization, and so on.[18] "Something" develops things every day that we humans are not able to produce, and this happens through the very same processes that have made us who we are.

In the context of the brutal climate changes to come, sites of free evolution are lessons in natural history: they teach us what a natural milieu in its full functionality is capable of doing. No one actually knows what a forest can do. Moreover, the effect on our ecosystems of a 2° Celsius temperature increase is also a complete mystery.

The decision to allow free evolution is thus a strategic decision in the face of the warming climate: at all events, milieus are going to face such colossal mutations that seeking to manage, administer, or control their evolution is going to become inconceivable on the practical level. And milieus that are not on life support, that do not require continuous interventions to be maintained, react better to such metamorphoses. They are healthy in Georges Canguilhem's sense:[19] they metabolize change better, they are more resilient.

*Anatomy of a Lever, a Case Study*

It is reasonable to think, in the face of uncertainty – and there is consensus on this point in conservation biology – that the more a milieu is simplified, impoverished, oligospecific, the more fragile it will be. And that the more a milieu is diversified, structured, rich in fluid and powerful functionalities – that is, the more its spontaneous ecological dynamics remain unimpeded – the more robust and resilient it will be.[20]

## *Hijacking the omnipotence of ownership*

Contemporary social science has shed a glaring light on the potential ecological and social violence of private property. The project of securing land control for hearths of free evolution entails turning the violence of ownership against exploitation at all costs, against extractivism.

But how can one be sure that this reorientation of property ownership does not become appropriation in the autocratic sense? Or privatization? What guarantees that the protectors of nature will not be corrupted by their own tool? The challenge is to figure out how to kidnap this legal arrangement while negating its risk of arbitrariness. How can the right of ownership be redirected while the toxicity of "absolute power" is simultaneously neutralized?

There are two answers to this question. The first lies in a public arrangement for clarifying, a priori, the way the milieu in question is to be used. A good example is found in the charter for wildlife reserves set up by ASPAS:[21] an association takes ownership of the property while tying its own hands in advance concerning the future of the space, through a public charter that describes its intended uses with precision. Such a charter could be adopted, moreover, by other private owners who want to restore their territories to free evolution,

## Rekindling Life

with the initiative of the "Hâvres de vie sauvage" project that ASPAS is currently developing.

The second response concerns the nature of the owner: nonprofit associations created under the French law of 1901, such as Forêts Sauvages or ASPAS.[22] As Philippe Falbet wrote in his response in the form of an open letter to the Confédération paysanne (a French agricultural union), which had accused ASPAS of seeking to "privatize" nature: "There is no question of 'privatizing' lands; these organizations are collective, associative, officially approved, originating from civil society. They are directed toward the general interest."[23] Associations such as ASPAS are not private actors defending private interests. They defend the interest of the living world in a period of major ecological crisis; consequently, they defend the public interest. Lobbyists acting on behalf of exploiters like to call associations for the protection of nature "lobbies": this is a strategic ruse for the purpose of placing them on the same level in the arena as themselves, knowing that in that arena their own economic and political power is far greater than that of the associations. But this assimilation of protectors of nature to a *lobby* is in fact a conceptual manipulation: the essence of a lobby is to work for *private interests*. The associations in question are working for the general interest, for a multi-species common good. They have no private interests to defend: life is not a multinational enterprise that pays them to traffic in the shadows in Brussels or elsewhere. They are the defenders of life.

Moreover, these non-governmental organizations are governed by the entire set of their supporters, who have banded together around a common project. Nonprofit organizations of this type have a specific political character: their most powerful organ is not the governing board, or the administration, but the general assembly, a collegial body consisting of all the organization's members. In the case of

*Anatomy of a Lever, a Case Study*

ASPAS, there are currently more than 13,000 members, and every donation for the Réserves de Vie Sauvage above and beyond the annual dues (€25, about $30) gives the donor the right to vote in the general assembly. These assemblies have the power to install and dismiss members of the board of directors. This arrangement neutralizes the risk of autocratic management by the directors: it fosters collegial behavior by members who have no private or exclusive rights with respect to the properties. This element constitutes the second guard rail that neutralizes the risks of ownership.

In this way, the right of ownership is reoriented but neutralized; in a powerful reversal, it is at the service of forms of life other than that of humans. We are entering into forms of self-management for life. We are working toward a common good – but the common is multi-species: it includes deer, silver firs, lichens, flowering prairies, and nitrifying bacteria.[24]

## *Owning in order to give back*

Classic ownership offers freedom to *take* something: the right to enjoy property, the right to use it for oneself. Here we are sketching in a right of ownership as a right to *give something back* – de-appropriation as the freedom to restore. It creates open territories. It offers itself to nonhuman and human users alike, rather than confiscating a space exclusively for the use of the owner. It entails a de-appropriation from within the right of ownership. This strange use of a right paradoxically makes it possible to de-incarcerate property from the movement tending toward enclosure. It is an improper use of the liberal right of ownership: a right of impropriety and a letting-go of land.

Add to that a campaign for citizen funding through an Internet platform of participatory financing, where all can contribute according to

*Rekindling Life*

their means to the collective acquisition, and you have the Vercors Vie Sauvage project: the associative acquisition of land consisting in a wildlife hearth destined for free evolution. Tomorrow's ancient forests will be "free for eternity," as Gilbert Cochet puts it. And the most reliable form of eternity we know in the liberal West is private property.

In a reversal of the relation between payment and enjoyment, here we have a commitment to pay in order to *leave* something in free evolution. The concept is paradoxical: agreeing as a group, in a citizen mobilization by way of a gift, to divert the absolute and exclusive right of ownership away from private enjoyment toward a radical restitution to other beings. This constitutes one response among others to the searing question raised by humans who know where they come from: how to give back, how to restore something to our giving environments? One could imagine acquiring a certificate (yet to be created) that would attest not that one owns a piece of property but that one belongs to a plot of land.[25]

We can quickly evoke the social effects of this type of project. The Deux Lacs reserve, created in 2013 by ASPAS, is an interesting example. Far from denying the local populations access to these protected spaces, the project goes out of its way to make the spaces available in a form different from that of economic exploitation – in the form, instead, of reconciliation and recognition. This is achieved through interventions in schools, through education in milieu protection at the local level, through displays in the communal media center or multimedia library, through initiatives addressed to citizens of the commune whose homes are adjacent to the reserve (by showing photos captured on an on-site camera installation), or through guided tours for small groups, led by a naturalist. The challenge is to relearn who inhabits these spaces along with us. To recall the richness of the living world that surrounds the local human inhabitants who have often forgotten about it or are uninterested. To reweave the bonds between them and their landscape so they can be proud of sheltering such a place.

The fact of leaving the forest in free evolution thus does not imply that humans cannot weave relations with forests. The project's detractors invariably respond that, if one cannot take anything, one cannot do anything. But this assimilation between "doing" and "taking" strikes me as a sign of a peculiar civilization. There are ways of "doing," in relation to a milieu, that do not imply extracting, exploiting, or taking away, and these approaches are no less real, no less powerful, no less serious. It is not the absence of any relation to a forest that is defended here, but a different relation – a reconnection. We must start by struggling against the extinction of experience. Among the myriad things at stake in these projects is a response to the absence of lived connections with forests. Without threatening free evolution, structures could be designed to welcome people who no longer have any familiarity with forests (ASPAS has already taken initiatives in this direction). One version would bring in small groups of schoolchildren, linking them with associations that are working on these issues: this would be a way of giving underserved urban populations access to nature. Such an endeavor entails a commitment to the right to nature, in the sense of a right to experience a forest that is allowed to develop on its own – an introduction of the populace to forests and to free evolution.

The limit of these initiatives of land acquisition for hearths of free evolution lies in their very strength: to what extent is the right to ownership on which these associations rely capable of conferring *enough time* to allow the protection to act on the scale of centuries? The big challenge is to make the free evolution of these little spaces permanent, on time scales that current institutions find hard to handle. According to some legal experts, the linking of property rights and the right to create associations constitutes one of the most effective arrangements in the current state of French law. But the proponents of free evolution projects are reflecting on more solid measures, in order to ensure something like protection in perpetuity. Must the project's defenders seek the status of "endowment funds," a status notoriously hijacked and subverted by the residents of Notre-Dame-des-Landes,[26] to assure the acquisition of a given property for the long run while sidestepping the classical forms of ownership? Or should they seek the status of "foundation," an institution whose principles are unchangeable, to neutralize the risk that, over a few decades, the governing board of the association in question could allow its management principles to stray far from free evolution? Should the defenders take inspiration from new legal measures, like the "real environmental obligation" of 2016, a form of servitude that makes it possible to leave a territory without an exploiter?[27] Or should they dig up the old legal principle of "unavailability," very present in the history of law, but more or less neglected today? Reflection on these questions is under way.

In a remarkably subtle text, Lionel Maurel has questioned the strengths and the paradoxes of these initiatives by ASPAS, starting from the legal and economic questions that arise about the status of a commons.[28] However, the either/or question "Should one work with or against private property?" seems to me badly formulated. The problem is not one of working with, or without, or against, private property. The problem is how to activate all possible levers simultaneously, in relation to precise, desynchronized agendas for struggle. Currently, it makes sense to "work with" the right of ownership, by diverting its toxicities as much as possible, and at the same time to carry out the legal work of creating statutes governing common spaces that could be substituted for ownership (here, we could have commons devoted to non-utilization, for example), and finally to undertake a political work of struggle against the multiple excesses of private ownership. These three projects are complementary, though they do not share the same time frames. Still, I do not believe that the projects of land acquisition for the purpose of protecting milieus are damned – in the sense of perverted – by the fact of using a law like that of private ownership.

To live and to struggle may mean to hijack. The endowment funds used by the ZAD of Notre-Dame-des-Landes are a product of the dominant extractivist economic system; they were not created for this use, but the residents diverted them from their original purpose. One can do the same thing for ownership. If lawmakers had already made available workable and effective tools for protecting these forests in the form of commons, it is probable that the initiatives for defending forests presented here would have willingly made use of them. In the meantime, it hardly seems appropriate to me to demand that activists renounce the legal tool of ownership on the grounds that it is part of the "system," when there is no equally powerful and durable alternative tool that could actually produce the results being sought, whose urgency is undeniable. That is, unless one complacently adopts the position of systemic critique and calls for a total revolution that will put an end to ownership, instead of taking concrete action on the ground.

# 3

# The Embers of Life

*From one cathedral to another*

We have seen how the Vercors Vie Sauvage initiative found the means to solve the problems posed by the unequal power relations between its activists and the exploiters' lobbyists, and also the problems involved in ensuring protection in perpetuity. Nevertheless, the central question remains: given the intensity of the current crisis facing the living world, is this type of action truly credible if only small plots of land are at stake?

The problematic aspect here is commensurability. Can an action of this sort exemplify an Archimedes lever for ecological actions on a broad scale? It could be argued that there is no common measure between this microscopic idea and the crisis we are facing: "We are in the process of destroying the living world, and you're talking about reconstituting confetti-like bits of forest. We're burning down the cathedral, and as a solution you're offering to restore the basin for holy water at the entrance." This objection actually reflects a profound philosophical misunderstanding of the nature of living beings.

In the wake of the fire that ravaged the great Notre-Dame cathedral in Paris in 2019, some troubling memes turned up on the Internet: photos of wrecked jungles, filthy beaches, dying forests, polluted sea-floors, with the same injunction under each one: "Rebuild this cathedral."

## The Embers of Life

Each of these powerful images sought to make visible the disproportion in financial means and mobilization between two instances of destruction of deeply precious things. And the comparisons make a strong point. But, like all metaphors, these images tend to block out other considerations. For life is not a cathedral in flames, it is a fire that is going out. Life is fire itself. A germinating fire. Here is another metaphor, dynamic, historicized, far from the imagined static image of a cathedral: it does more justice to the originality of the phenomenon of life.

To avoid falling into the trap of analogies, it is important to begin by spelling out the limitations of the metaphor of a cathedral in flames when it comes to describing the status of biodiversity. The analogy with the destruction of a cultural monument implies the idea of definitive demolition. But life, from the evolutionary and ecological standpoint, must not be conceptualized initially as a monument razed by "barbarians," like the Greco-Roman temple of Baalshamin in Palmyra blown up by ISIS. Life cannot be compared to a monument built by human hands, or to a tribal language lost forever. Life is not a patrimony in the human sense – something made by humans, fixed and fragile, inflammable: it is, above all, a creative fire. We did not make it; it has made us.[1]

By the metaphor of fire, here, I mean that, while the biosphere can indeed be reduced, impoverished, weakened, it takes only a few embers and a lifting of constraints (liberated ecological niches, more clement conditions) for life to proliferate, multiply and spread in all directions. The absence of life in a particular place reflects only its stubborn, relentless obstruction by conditions external to it, challenging its vital, tireless abundance.

By "creative fire," I mean that this vital radiation can always invent countless new forms. The original force of life, which has been expressing itself during more than 3 billion years of evolution, can be summed up in a word: prodigality.[2] The "absurd extravagance" of life,[3] its jungle-like power to regrow, its tendency to overflow in torrents, are such that if the conditions for life become favorable again, from an ember, from a small milieu (as long as it is ecologically vivacious enough), a flourishing life can be reborn, capable of major evolutionary radiations. But, for that, the last remaining embers must be cherished. And not in the fantasmatic form of specimens in zoos: rather, in the form of living animal and vegetal populations, in protected and integrated milieus (for the habitat of one life form is nothing but the weaving together of all the others). These milieus require great connectivity, and populations in adequate numbers to ensure genetic robustness and a capacity to change, if they are to adapt to the environmental metamorphoses that necessarily follow in the wake of climate warming.

Darwin had a powerful intuition of the pyric nature of life when he modeled it by means of a thought experiment that played a crucial role in his understanding of evolution. In a book on orchids published in 1862,[4] his first after *On the Origin of Species*, he noted that certain small orchids have as many 180,000 seeds per plant. He calculated that if all these seeds succeeded in sprouting, they would cover about an acre; at this growth rate, their great-grandchildren would cover the earth. In other words, a single orchid would spread over all the land on earth in four generations, thus in four years – if, and only if, all its descendants lived. They would all be different; each would be unique. And we need to understand that, *virtually*, each of the 10 million species that inhabit the earth has an *analogous* power, adjusted for growth rates. The relative stability of populations comes from the fact that the earth's inhabitants are already engaged in activity, and that all of them want to live. But if room is left for it, and as soon as conditions allow, the intrinsic nature of life will express itself: this is the creative proliferation of variations, capable of covering the world like fire.[5]

To understand the evolutionary nature of the biosphere, we need to conceptualize it, then, as a living fire, a *prodigal fire* – this said without the slightest mysticism, unless it is of a tranquil sort, required by the spectacle of the eco-evolution taking place outside of us and within us.

The naturalist Gilbert Cochet offers one concrete example, among many others.

> In 1998 the Maisons-Rouges dam on the Vienne River was taken down. A boulevard was thus opened up for migrating fish. In the ensuing months, sea lampreys and shad came back to reproduce where stagnant, blighted water had reigned for three quarters of a century! Year after year, the numbers have increased, probably reaching some 100,000 individuals. What did we do to achieve such good fortune? Nothing. We simply undid the infamous concrete obstacle that had sterilized the Vienne basin for too long. The point is worth emphasizing: no pisciculture was involved! The lamprey and shad came back on their own, without our help but also without our hindrance.[6]

Is it a matter, then, of restoring ecosystems, the way one restores a famous painting, or a church? In other words, do we apply our organizational genius to a fixed material object in order to restore it to its original state, to fight against the "march of time," a figure for the entropy that necessarily damages the entity in question? Importing the "restoration" metaphor drawn from patrimony made by human hands into ecological engineering reveals our profound misunderstanding of the living world and our relation to it. In the living world, the march of time is not entropy, and there is no original model to be sought: the flow of becoming creates and recreates forms, organizing itself on its own. In it we can restore nothing: the dynamics of life alone are capable of restoring themselves. We can at best restore the minimal conditions for this restoration to occur. In other words, pushing to the outer limits the paradoxes of the technical metaphor applied to ecological engineering, we can at the very most act with care and restraint to repair mechanisms that we have broken, in a machine that we have *not constructed*, so as to let it reconstruct itself on its own.[7]

In short, we may be able to restore what we ourselves have made; but we cannot restore what has made us.

Marine biologists have recently made a discovery that bears a powerful philosophical message: the population of humpback whales, those fabulous cetaceans that can live as long as 80 years, seems to have returned to its original size – that is, as it was before the extractivist years of whale hunting. Did we restore the whales the way Versailles was restored? Clearly not. What happened, then? All we needed to do was set up the right protective measure for a long enough time; here, this simply meant putting an end to persistence hunting, prohibiting all forms of whale hunting from the late 1970s on. And in 40 years, the population has irradiated into a bouquet of life: from around 440 individuals in the late 1950s, at least 40,000 seem to be traversing the oceans today.[8] We have simply let the dynamics of the population do their work. Here is another magnificent example of the pyric nature of life. We do not regenerate life, we prime its autonomous powers of regeneration: we let it express its own resilience; we put in place the minimal, delicate, discrete conditions that will allow it to regain its full vitality.[9] We ensure the conditions for autonomous repair of the arrangements of autonomy so we can disappear as repairers. To defend the living world is in one particular respect comparable to bringing up a child: it is a matter of working toward one's own uselessness as educator or developer – working toward one's own self-effacement.

## *Defending the embers*

The idea of defending hearths of free evolution is commensurate, in a way, with the scope of the current crisis of the living world, because it is betting on the same property of life. Life is not a cathedral but a fire – a fire that reconstitutes itself all by itself, deploys itself, creates countless forms, as soon as we leave it the space and the time.[10] And this is not a coincidence, for those who have come up with the notion of free evolution are first-rate naturalists who have carefully studied the power of milieus to reconstitute themselves as soon as the pressure on them is relieved: the return of pollinators when pesticides are banned; the return of certain species of migrating fish toward the sources of our rivers when dams are removed; the return of predators to forests rich in game as soon as we stop exterminating them. And we do not necessarily need large spaces to achieve this (even though the ecological effect on large spaces would be far greater): naturalists have shown that even a small island of senescence constituted by a single venerable tree, or a few dozen acres of forest, can play the role of spreader of life, can serve as intermediary for increased regeneration, if we allow it enough time.

The paradox that our mammalian species does not fully grasp, being a species for which aging implies decline, is that forests do not age the way we do: in aging, they shine and spread. Old forests are fountains of youth. The more we let a forest age, the more it rejuvenates, the more it gains in the power to reinvigorate the world around it.

## Rekindling Life

If the living world were first and foremost a cathedral, the war would already be lost. If life is fire, the problem is a different one. It is on a scale we can handle, if we only give ourselves the necessary levers, awareness, and motivation. The problem then becomes, above all, the problem of how to protect the embers. How to defend the embers of life all around us?

This is our fight. We shall struggle in the forests and the mountains, in our gardens and our cities, in the fields and on our streets: we shall never give up.

What is at stake is maintaining and recreating the conditions that will allow these embers to blaze up again, restoring healthy habitats and milieus without destructive chemical interventions, reinstating genetically connected populations, unfragmented habitats, hearths of free evolution, connecting corridors, and so on.

If action is possible, it is because the original power of the living world, dating back billions of years, lies in its abundance of vital propositions, its generosity in gifts, its multiplication of differences: the biosphere is a living fire that covers the earth, and it can always start up again if we know how to defend and kindle its embers.

In the light of this metaphor, what becomes of the preserves given over to the other forms of life, the wild forests, the protected zones: all those *hearths* of free evolution? By now, it should be clear why I chose the term "hearth" to unify this idea: these spaces are hearths that protect the embers of life and radiate outward. They are living, shared hearths, teeming with life: "hearths," since it is from *there* that everything can start up again. They are open hearths (we can go in, other living beings can go out) in which we jealously protect the embers, multiple origins for the departure of flames to come. Hearths of living resistance, like backfires countering the productivist war being waged on milieus. They are there to reinvigorate the embers of life and to maintain its potentials for tomorrow, so that our damaged world will be set ablaze with new life.

Life as fire is a circumstantial metaphor, relevant with respect to a particular complex of problems, and not a descriptive concept purporting to be exhaustive.[11] Certain aspects of the way life works are well translated by what the human imagination associates with fire – its fundamentally dynamic dimension, rich in almost immaterial properties, for example: the evolutionary and ecological dynamics are in fact the essence of the living world, the coevolution of memories that each living being constitutes and transmits, and not the stationary biomass in some particular ecological compartment, for the biomass is in constant circulation. From the standpoint of action, this metaphor makes it clear that it is not, strictly speaking, the biomass that must be protected in a forest, but rather the dynamics themselves, the memories, the equilibrating arrangements and the adaptive potentials.

Thus, the evolutionary function of life is also preserved in the metamorphic imaginary of flames: the variation and the evolutionary transformation of forms that radiate when niches are available.

The metaphor is limited, but it warms and enlightens a little, allowing us to pause, to regenerate our energy and set off again to invent better formulations, actions that are more fitting and better adjusted to this world.

## *The new war of fire*

The story of the ancient war of fire is that of mobilization, attention on the part of the whole tribe directed toward protecting the embers on a daily basis. To keep them from going out altogether in periods of wildness. In a sense, what is at stake is still the same: it is the nature of the fire that has changed. The challenge today is to defend the embers of life, to cherish and rekindle them so that they will start back up on their own and once again spread their warmth, their light, their generosity. To protect them in a thousand ways: a hearth of free evolution is just one way among countless others to be invented.

This is the new "quest for fire":[12] from now on, the embers of life are what must be protected. And this time it entails war against ourselves. But we must be careful. This is not a war against ourselves as a species. Not as a totality. And it is not the necessary fate of humanity. As I see it, we need to criticize the latent misanthropy of certain visions of the protection of nature, visions that circulate in some of the associations mentioned in this study. Humans "in general" are not the problem; rather, it is the slide of a recent economic and political form, a social metabolism running wild, a particular relation to the world that has been set up as a norm and as Progress writ large: something like a financialized productivist extractivism, extending market logic to everything that ought to be excluded from it, and incapable of any sobriety. This is backed up by a modern culture of "cheapening" life: in other words, by a process that simultaneously devalues life ontologically, depoliticizes it, and converts it into inexpensive raw material.[13] But humans are also the solution to the problem raised by certain human activities and their systematic logics.

Under our circumstances in this third millennium, the new "quest for fire" picks up the old myth of the human project in its relation to "nature" and subverts it. The Moderns have thought that they were engaged in the project of dominating and conquering the web of life to the benefit of human society, which had extracted itself from the interwoven biotic communities. But this vision is in fact provincial, and recent: the journey of humans over 300,000 years, in their relations with the living world, has been – under countless irreducible cultural visages and under all latitudes – a project of habitability. A project not of appropriating the world but of adapting *to* the world. The fire used by our ancestors, starting some 500,000 years ago, is first and foremost a source of life, not a weapon of destruction. This human journey is indescribable, but in a sense it can be summed up in a phrase: making life livable and the world inhabitable.[14] Now, the recent conscious discovery of ecological thinking, a discovery that some peoples are already putting to work every day in their relations to the living world, is that life is livable for us only if it is also livable for the other living beings, since we are only a knot of relations woven together with the other forms of life.

If this somewhat dated motif of the quest for fire is interesting today, it is because it reminds us of the necessary link between the future of humanity and protection of the embers of life – it reminds us that the two are inseparable. As in the old story of the quest for fire, the tribe cannot be saved without a collective commitment to the embers – its members must stand watch together over the embers of life. Most importantly, that collective commitment resembles less an abstract, voluntarist slogan than a protocol for specific action: it is more relevant, more to scale, more feasible, to protect the embers than to plunge into a burning cathedral with a bucket of water, or to panic and pray on the forecourt in the face of our presumed impotence.

## *Beyond protection*

The metaphor of a new quest for fire, finally, activates the often overlooked twofold idea that we do need to protect life but, paradoxically, as something stronger than we are, older than we are. It is a force, a web, a process that has made us. In the Western cultural tradition, our prejudice holds that what we must protect is always weaker than we are, more vulnerable, something *minor*. In *Beyond Nature and Culture*, Philippe Descola has shown that the relational schema of "protection" is by definition asymmetrical – it necessarily implies that the entity protected is *inferior* with respect to the protector.[15] This relationship model is a signature of naturalism in Descola's sense: conceiving of care for the milieu in the form of "protection" is a strange view provided by those who have invented "nature" as other, nature as opposed to humans deemed superior owing to their exclusive possession of rational interiority.[16]

The old "protection of nature" has to be rethought in terms of a *defense of the living world* precisely because the latter is in no way inferior to us. "Protecting nature" is paradoxical in a way that is rarely brought to light: it means protecting what is more powerful than ourselves. It means watching over the embers of a prodigious fire. An ambivalent fire, for, like fire itself, life is not here "for" us – it is not benevolent; it is generosity incarnate, but it can also be dangerous; we have to negotiate with it.

And, indeed, the metaphor of defending *fire* stimulates the protective passions, but we are far removed from the strange megalomania of compassionate attitudes toward "nature," attitudes that represent the protection of life according to the model of protecting vulnerable beings: children, the handicapped, other defenseless beings for whom we are responsible. The metaphor of fire restores to life the paradoxes that connect us to it: fire is not in our power, but it is to be defended; it is our world, but it is also those uncertain embers; it is weakened by our infringements, but it is more powerful than we are; it calls for our protection, but it is much bigger than we are. To defend it is to kindle it. Moreover, this is why the only relevant forms of protection consist in wagering on the power of fire – the power of life – itself: we have

## The Embers of Life

neither the means nor the competence to "save" life by technological interventions, the artificial design of a new earth ecosystem, pollination by drones, or any other form of geo-engineering. Inasmuch as its powers surpass ours, life is the *only* thing that can regenerate itself, if we let it do so – if we restore to it the conditions allowing it to express its resilience and its inherent prodigality.[17] If we kindle its intimate strengths, simply by helping to reconstitute the conditions under which it can express its own cascades of energy, cycles of matter, potentials for evolution. If we protect what surpasses us and includes us.

In certain respects, we have chosen the wrong vocabulary for conceptualizing the traditional "protection of nature." That concept was dualist and patrimonial: both "protection" and "nature" have to be rethought. In the context we are concerned with today, we need to change the conceptual framework of our imagination: protecting nature does not mean adopting an overseer's posture and taking charge of something other, of an outside conceived as a vulnerable, passive, impotent temple – it means kindling the embers of a multiform fire, the one that constitutes us, the fire of which we are just one face, the fire that builds and rebuilds itself ceaselessly through its own power, sheltering us and giving us life in the process.

The idea of "protecting nature" contains an additional pitfall: it convokes "nature" as an entity inherited from the modern, dualist cosmos that divides the world into two separate blocs – humans on one side, and "nature" on the other. What becomes of "protecting nature," then, once we have understood that the word "nature" has sent us into a dualist impasse, and that "protecting" was a paternalistic way of conceptualizing our relations to our milieus? It becomes "rekindling the embers of life" – that is, struggling to restore vitality and full expression to the dynamics of eco-evolution. It becomes defending our interspecies milieus of life: these milieus are forces that constitute us, forces that are greater than we are but that nevertheless require our care.

To speak of rekindling the embers of life is simply to redescribe what used to be called "protecting nature," before we moved beyond belief in "nature," beyond the paternalism of protection and beyond dualism. It is a formula that brings together three philosophical propositions: life is not a cathedral, but a fire; while we cannot protect something that is greater than we are, in a paternalistic manner, we can restore to it the conditions of its own autonomous regeneration; it is not as humans that we protect an other that would be "nature," it is as living beings that we are defending life – that is, our multispecies milieus of life.

It is in this sense, finally, that it is important not to conclude that defending hearths of free evolution is good only for the "greens": it is not primarily in the name of a love of flowers and animals that it makes sense to protect these spaces, it is in the name of protecting the constitutive relations that weave us together with the world – in the name of the imperative, clear as water streaming from a rock, to defend the world that makes us. It is a matter of defending the living world not because that world is useful for us in terms of quantifiable services, nor because it is vulnerable and calls for our compassion: rather, we must defend it for its very powers, those powers that have molded us, woven us together with all the other forms of life, and that still infuse us with life every day.

## *The last impropriety*

In this light, the discourse of compassion toward nature in general has a preposterous aspect: it reveals its ignorance of what it claims to cherish. The living world is our "giving environment"[18] – it is the instruments that sculpt us, nourish us, mistreat us, give us joy; it swarms with powers with which we can negotiate; it is not a baby seal assaulted on the Internet, whose face stimulates the instincts of tenderness and pity. This latter is a bias of the late Moderns, who live in a world *made* by human hands, and take as their model of "nature" a beaten or slaughtered animal, or a victimized polar bear.

The gorges of the Lyonne river, which harbor the Vercors Vie Sauvage reserve, teach us the impropriety of the one-dimensional attitude that conceives of life essentially in terms of its *vulnerability* and thereby reduces the gamut of relations that we might have with it to compassion from on high. When one enters deep into the reserve for a day, scratched by brush, stunned by myriad bird songs, immersed in the orgiastic sexuality of pollens, abuzz with a thousand industrious and cosmopolitan life forms, living *in the minority*, under the amorous displays of eagles, we sense intimately how inappropriate it is to speak of "nature" as a fragile little thing that has to be "saved." We suddenly perceive the obscene myth of omnipotence that is papered over paradoxically by compassion (hell is paved with good intentions). The living world is our world: it is an interweaving of life forms that command respect. Its powers call for incessant negotiations, so that we can live from them and live with them. These are thick powers of time alive with ancestralities, to be translated, influenced, composed, despite their reticence, in order to invent a cosmopolitan milieu. But this is not a last defenseless innocent to be saved, it is a prodigious fire to be rekindled, a fire to be defended. To protect this world is not to save innocents; we are life defending itself.

## Rekindling Life

### Far from collapse

What is required intellectually and politically here is to hold together the two dimensions of a seeming paradox: the *reality* of the ecological crisis (one out of every eight species may disappear in the coming decades) and the *inherent prodigality* of life (from a few dozen surviving beavers in France in the mid twentieth century, there are more than 50,000 today, owing simply to the relaxing of constraints on their habitat, along with intentional policies of protection).

To put it bluntly, the biosphere is not about to disappear.[19] To affirm that it is, as some complacent commentators do, sincerely believing that they are defending the cause of life, that "life on earth is collapsing," is at best vague (what does collapse mean, precisely?), at worst completely wrong and gratuitously catastrophist (if we take the metaphor seriously and imagine the collapse of life to be like that of a building, or a system, the image is not applicable, for the great ecologic and evolutionary functionalities are not on the verge of collapse).

In fact, the living world is not "collapsing": the vagueness of this formula, its apocalyptic aura, do not do justice to life (nor to us: there is still some anthropocentric megalomania in this business). No, the biosphere will not "die": what is in danger are countless living forms and relations among living beings, timeless interweavings, and finally our constitutive relations with the actual living world (and not life in itself) that risk disappearing.

We need to do justice to the paradoxical character of the situation: we need a good understanding of what we are seeking to defend, so as not to fall back complacently into caricatural and self-fulfilling apocalypses (if "everything is fucked," our protective energy starts to diminish right away, and we don't even know

## The Embers of Life

where to apply the energy we have in order to change direction). It is important to learn to maintain empirical probity in the effort to understand what is happening, in the face of over-the-top prophecies: we need to learn to keep two seemingly contradictory propositions in our minds at the same time. For example, in Western Europe, it is a question of acknowledging the catastrophic character of the loss of small fauna (first and foremost, insects and field birds) even as we recognize the return and the prodigious reinvigoration of large fauna in once-deserted landscapes (the return of vultures and ibexes thanks to the nudge of reintroduction, the increase in large and small deer, the return of salmon in the Allier river, and dozens of other instances).[20] To be sure, from the standpoint of the functionalities of the various milieus, what is disappearing is more significant than what is coming back – but this points to the real enemy: in the case in point, the massive use of phytosanitary products and, more broadly, agrobusiness and its "world" are what emerge from this diagnosis as the principal responsible parties.

The problem needs to be formulated with care and precision, to avoid instilling the megalomania of progressivism into the very *critique* of the progressivism of techno-industrial modernity. We are not putting the living world in danger – but what is happening is no less serious or tragic for all that, and it does not permit us to carry on with business as usual for an instant. What we are putting in danger are untold numbers of living forms, swatches of diversity, and ultimately our constitutive relations with the living world that maintains us – thus our own conditions of life, along with the features of the species and ecosystems that have shared the evolutionary adventure with us over these last hundreds of thousands of years. And this is certainly enough to require a complete change in our relation to the world, to production, to exploitation – in short, in our relation to life.

## A lever of ecological action

Of course, the project of creating hearths of free evolution will not "save the world" all by itself, but it is a good example of the type of ideas with powerful hands of which myriads more must be invented to save what must be saved. And it can serve as a banner for a new "quest for fire" whose forms will necessarily be multiple.

The project of creating hearths of free evolution by acquiring real estate is interesting because it brings together certain features that are highly desirable today for a lever of ecological action commensurate with the crisis. The Extinction Rebellion movement, for example, inspires citizens looking for radical action, citizens rightly revolted by the infinite compromises of our politics. Here is radicality: no more compromises with the exploiters, who are omnipresent in the management of public territories. No more power granted to the agrobusiness or hunting lobbies; for once, they are kept out – an achievement on the part of those who defend the fire.

Are we looking for effectiveness? Here is an example: let us contribute to tomorrow's old-growth forests, which are among the best carbon sinks on the planet. Do we want to see a subversive shift of the reign of the financial world to the benefit of the living world? Do we want effects that are perceptible right now, a fight on the human scale, and the fruits of action amplified to the scale of potential old-growth forests in the future, in 800 years, 8,000 years? Hearths of free evolution offer a way forward for the guardians of the fire.

But there are other paths to open up. This is not a book about free evolution, it is a book about levers. What interests me in the particular case I have chosen to examine is how a seemingly innocent idea develops as it is tweaked by an encounter with heterogeneous ideas, still implying a sense of history, a relation to life (adjusted consideration),[21] compatibility with a "world" in the sense of a global project for society, and maturity in political ecology (no solutionism) – all of which, taken together, have the effect of *reclaiming* the defense of the web of life, reappropriation by the citizenry of care for our milieus of life. But citizen reappropriation does not mean disengagement on the part of the state. The state must strengthen its role as defender of the living world; it must support and energize citizen initiatives, rather than dropping out of the game.

These levers for defending forests must take several forms, from free evolution to nonviolent sylviculture, with countless variations in between. These already abound in France where free evolution is concerned: PRELE, FRENE, RAF.[22] Even the French National Forestry Bureau (ONF) is taking timid steps in the direction of free evolution. But action is also taking place at the level of collective, non-institutional initiatives: in Haute-Savoie, a community of villages has come together to buy a forest; its leaders are working out arrangements for making the site a commons, and they are reflecting on the adjusted consideration required for defending the dynamics of life.[23] Here is an example of citizens reappropriating the defense of the living world.

These are lightning-fast combinations of specific ideas that, taken together, make it possible to shift the world on its axis in order to make it slip into another valley, a path of least resistance, in harmony with life.

## Rekindling Life

I cannot do justice here to all the levers that have been invented everywhere – there are thousands buzzing around on the ground; the inventiveness of actors plugged into a living territory they care about is inexhaustible. Let me simply isolate some invariants of these levers, so that they can be recognized more easily.

The idea that can become a lever is an idea anchored in the viewpoint of interdependencies.[24]

It is an idea that, even if it is local, has to fight for "a world," as the Notre-Dame-des-Landes defenders were fighting "against the airport and its world" – that is, the idea must contain the seed of a project for society.

But this does not mean that it is innocent in all respects, compatible only with a demanding ideal that requires us to wait for the big day.

The idea has to rekindle the embers of life. It has to be situated above the dualism that separates humans from "nature." It has to take its place in a culture of struggles for the living world. Once it is found, all we have to do is invest all our collective energy in it. Honestly, what better use could we make of our time?

### Exiting the divide

When we look at nature reserves as closed in on themselves, without links to the world around them, we run the risk of defending territories untethered to the earth. When we succumb to this view, the free evolution initiative seems to manifest lack of interest in the rest of humanized territory, which makes its defense of wild forests look like a fight for the benefit of "nature" *alone*. Since we have unfortunately inherited a *dualist* conception of the relation between "humans" and "nature," in the collective unconscious, fighting for nature seems to be undertaken *against* humans, to their detriment. But with the reserves, the interests of living beings, humans included, are indistinguishable: a logic of relations is required. This logic forces us to leave behind the dualist logic in which the good of the one is necessarily achieved at the expense of the other.

The initiative in question is not beneficial to nature at the expense of humans: its benefits are not reserved for "nature" inasmuch as it serves the well-being and the survival of humans. It is a way of acting for the good of the indivisible community of living beings, a community to which humans belong.

Consequently, an important stake lies in the capacity to connect hearths of free evolution with the rest of the territory: they have to be conceptualized in an integrated vision of the forest.

We can anticipate the lines of force of this pluralism. Everyone is prepared to feel that a nonviolent form of forestry dedicated to producing wood, even while respecting forest dynamics, lies, in a sense, much more in the camp of the defenders of a forest in free evolution than in that of the exploiters of a forest in monoculture, who are skilled at clear-cutting, planting, and chemical interventions. Nevertheless, it is as though, prisoners of dualisms – trapped in the opposition between exploiting and sanctuarizing, between productive nature and nature left to itself – we have no concepts that would allow us to express in depth the nature of this alliance. The alliance to be conceptualized in fact binds together a strong defense of milieus that forbids their exploitation with certain forms of exploitation that are capable of protecting milieus. It is in the reduction of these paradoxes that we find the possibility of envisioning, formulating, and activating the alliances we need in order to defend the web of the living world of which we are singular components.

*Rekindling Life*

How can we shift over onto another map of the world, one that would allow us to think spontaneously in these terms – and breathe outside of dualism? What new cartographies can we devise so that borders no longer pass between us and our false enemies, so that the real alliances are made visible, the front lines made effective for the defense of the web of the living world – so that the cause of humans is no longer opposed to that of what used to be called "nature"? How can we conceptualize a defense of our living world beyond nature and culture? This is the object of our investigation from here on out.

4

# Realigning Alliances

*Exploit or sanctuarize? The impasses of dualism*

Where the uses of the earth are concerned, we have inherited a dualist cosmology, formulated as a fundamental opposition between exploiting and sanctuarizing – as if these two uses of the earth were mutually exclusive, conflicting, and irreconcilable. This philosophical opposition has found direct expression in local political struggles in France, for example in a motion adopted in 2019 by the Confédération paysanne de la Drôme (Drôme Farmers' Confederation)[1] "against the rewilding and appropriation of lands": the statement squarely targeted ASPAS's plans to create wildlife reserves. In this text, the Confederation claimed that free evolution, forbidding all exploitation of the forest lands ASPAS had acquired, was inimical to local agriculture, even of the intensely ecological variety, inasmuch as free evolution – starting from a project based on a dualist zoning of the spaces in question – appropriates lands that ought to belong to farmers. According to the Confederation, the protectors of nature appeared to assert, with their free evolution project, that any exploitation of the land, even by farmers, necessarily destroys forest milieus, and that only free evolution, with its rejection of all exploitation, constitutes a good use of the earth.

But such discourse, despite the intelligence of the people involved, has created a baseless opposition, because the words and concepts available to the various actors – those who defend local farming and those who defend strong protection of the land – are prisoners of the dualist metaphysics. It is as though ASPAS and Forêts Sauvages have not had the vocabulary to defend forests from any standpoint other than that of *dualism*, which implies that exploitation must be demonized if one is to defend the creation of sanctuaries. It is as though the actors of the Farmers' Confederation have not had the words or concepts to defend and valorize sustainable, farm-based, and intensely ecological exploitation of their milieus without condemning sanctuarization.

Why this misunderstanding? It has deep roots: essentially philosophical in nature, it is hard to see because its source lies in an ancient heritage dating from the very foundations of our culture. The misunderstanding stems from the fact that the modern West has a *hierarchical dualist* heritage – that is, a way of conceptualizing the world in terms of binary opposites, mutually exclusive and differentially valued: for example, "humans" versus "nature." The dualism functions in our minds through a law of inverse proportions, following the zero-sum principle of communicating vessels: what is good for the one is bad for the other, what elevates the one lowers the other, what is given to the one is *taken* from the other. When proponents of rewilding claim to be defending "nature," this dualist legacy comes into play, and many people unconsciously hear that "nature" is being defended *to the detriment* of humans, *against* human interests, since, for dualists, what helps one side always harms the other. Owing to this specter of hierarchical dualism that haunts us, the human pole of the opposition finds itself devalorized simply because what is wild is being revalorized. The problem for heirs of dualism is that they are always obliged to sacrifice and devalue *half* of the world in order to defend the other half. This is a wearying legacy.

*Realigning Alliances*

The language of dualism thus creates an irreconcilable conflict, multiplying unnecessary enemies and disorienting courses of action. For, as I see it, farmers and defenders of free evolution, opposed in their discourse, are *both right* from the standpoint of their practices and actions. (This is not to say, in an effort to find consensus, that *everyone is right* somehow, for certain parties are unequivocally wrong – for example, the defenders of agrobusiness, insofar as it is unsustainable for the milieus and alienating for farmers.)

In order to develop a tool kit of terms and concepts that do not create unnecessary enemies, it is imperative to think in terms other than those of dualist oppositions. What I am seeking to sketch out here are maps for orienting ourselves differently.

The somehow unreal character of the dualist dichotomies, despite their persistence and their omnipresence, becomes apparent in an intriguing way once we explore other technological initiatives concerning uses of the milieus we call forests. Let us look, for example, at the Réseau pour les alternatives forestières (RAF, Network for Forest Alternatives), which defends a type of forest management based on a sustainable, nonviolent form of sylviculture, attentive to the regeneration, richness, and resilience of forests. The RAF clearly belongs to the exploiters' camp. In the framework of dualism, then, it ought to condemn sanctuarization. When we read its proposals, though, we begin to appreciate the fact that its members are thinking and acting spontaneously beyond the either/or choice between exploiting and sanctuarizing. Indeed, the RAF proposes that, in each section of a forest exploited according to its recommendations, 10 percent be deliberately left free of all exploitation – in other words, in free evolution, following ASPAS's norms for wildlife reserves (which the network acknowledges as a source of inspiration).[2] And it specifies that this 10 percent must not be the areas that are the hardest to exploit, those with the least potential economic value: they must be the most interesting 10 percent from an ecological standpoint — that is, from the standpoint of what the *forest itself* requires.

Here are forest managers — that is, exploiters — who are spontaneously defending — and integrating into their management — the need to sanctuarize a portion of their resources. What has happened? How have they escaped from the dualist opposition? What conceptual dodge, what emancipating intuition has allowed them to bypass our metaphysical legacy so readily? It is certainly not that they want to satisfy everyone, through a feeling of ecumenical consensus or a pacifying synthesis; they are not afraid of enemies. In the deepest sense, the RAF and its initiative can be said to have originated in a *struggle against* the dominant actors in forest management: intensive industrial sylviculture.

To conceptualize a political surpassing of the opposition between exploiting and sanctuarizing, then, we might turn to the tentacular enemy that is *common* to the Drôme Farmers' Confederation and the associations for protecting nature that defend free evolution. This enemy consists in all the unsustainable, extractivist, destructive uses of milieus, alienating for workers, that are embodied in intensive, industrial, interventionist agrobusinesses. Let us call this enemy the "camp of unsustainable exploitation." This alone would be enough to motivate an alliance.

But the fact of having a common enemy — namely, a powerful anti-ecologist lobby — does not suffice to unify the various users of the earth who intend to care for it. We must reach the point of conceiving of a spectrum of relationships with the living world that have more than a shared enemy — relationships that have something deep and real in common.

## Realigning Alliances

To open up the path to this common denominator, we need to begin by leaving behind the idea, prevalent among protectors of nature, that all exploitation is destructive. This is a collateral effect of our dualist legacy. All forms of exploitation *modify* the milieu, but some do so in a way that is sustainable, and even sometimes life-enhancing for biodiversity, in a given context.

But it is extremely hard to navigate in the waters of dualism, since, to get away from that murky space, we must always *simultaneously deconstruct the two opposing myths*. In other words, we must reject at the same time both the idea that all exploitation *destroys* the milieu and the idea that all exploitation *improves* it. This latter idea may seem counter-intuitive and easy to challenge, at a point when the disastrous effects on their respective milieus of intensive agriculture, extraction of fossil fuel, and industrial forest management are so well documented. But its roots go so deep, it holds such an architectonic place in the relations of the Moderns to "nature," that it has to be formulated clearly, in order to exorcize the specter that sneaks in through the window opened up by the most innocent words, even when we have thrown the notion out through the door of ecological and historical analysis.

### Deconstructing the myth of improvement

The imaginary construct that imperceptibly structures our modern way of thinking about good and bad uses of the earth has to do with *enhancing the value* of nature. According to this conception, which was crystallized in England under the Stuarts, human action is necessary to fulfill, improve, and ultimately attribute value to a wild nature that would otherwise be deficient.[3] Ever since, modernity has been in thrall to the idea that wild milieus (what we are calling here areas in free evolution) are deficient, in the sense that they lack the human interventions that alone can fulfill them and make them valuable. According to the Moderns' foundational myth, their manifest destiny is to *improve* their milieu – that is, to ameliorate it and thereby enhance its value. In the context of the Baconian scientific revolution, the motif of *improvement*[4] renewed and enacted the very old Judeo-Christian motif of "stewardship," according to which the earth has been entrusted to humans: we are charged with enhancing its value.[5] And failing to act as a steward is a crime (a view that justifies colonial expropriation of autochthonous hunter-gatherer peoples, for, according to this ideology, they do not "produce" but are content to be "predators," a designation that associates them with the stigmatized carnivores: "civilizing" them amounts to transforming these "savages" into humans).

The idea of "improvement" posits that exploitation, as a way of putting milieus to work under the guidance of human reason, *necessarily* ameliorates them, thus fulfilling the destiny of milieus that, left to themselves, would be in a state of neglect. J. Baird Callicott synthesizes the postulate this way: it is "as if nature were created unruly and were in need of breaking to become complete."[6]

The imperative to *improve* is based on a fundamental confusion. If human reason and labor can improve the yield of a milieu in terms of commodities consumable by humans (which is true), improvement *for this purpose* is not necessarily improvement *in itself*, the correction of an intrinsic defect in the milieu.[7] Thus, it is not the idea of improvement that I am critiquing here (it is an agronomic fact that labor improves the productivity of an agrosystem); I am criticizing its acceptance as an absolute principle. Improvement indeed enhances the value of a milieu *in the economic sense* of "value"; it ameliorates *from the standpoint of yield* in terms of consumable, and presumably marketable, products. However, obscuring the latter part of these formulas to assert that the process improves and enhances the land, period, has absolutism as its principal metaphysical consequence – a disastrous legacy. It postulates that ecosystems that are not enhanced by human reason and labor are incomplete in themselves, defective, in want of our action. Taking improvement to be an absolute principle is a surreptitious logical implication that is nevertheless necessary for the doctrine of stewardship. To the extent that the God of a monotheistic universe shaped the Creation for our use, improving milieus *for our benefit* amounts to improving it *in itself*. The devil is in the metaphysical details.

It thus becomes apparent that we need to move beyond this erroneous conception of the living world by affirming that milieus in free evolution are not defective. The ecosystems harbored by the earth from the origin of life onward are tirelessly woven by the dynamics of living beings: coevolutions, diversification, successive generations in forests, journeys of sediments and fish in rivers, pollination, creation of humus by soil fauna, circulation of matter and energy in trophic networks. These dynamics of life cannot be called deficient, since, rigorously speaking, it is they that have shaped us, body and soul – they that have invented all the ecosystems, all the species of plants, animals, bacteria, and fungi that keep us alive. These are the dynamics that have created the world that welcomes, shelters, and feeds us; they have created us adjusted to this world. Consequently, any philosophy of life that is not oblivious to our origin has to assert in non-negotiable terms that the dynamics of life do not *need* us, must assert that these dynamics are neither deficient nor defective without us. We have only been here for some 300,000 years at best, we have only been putting the living world to work for a few thousand years, and we have only been improving it "rationally" for a few centuries or decades – while life, for its part, has been weaving its prodigies for almost 4 billion years.

This mad absolutizing of the idea of improvement, inherited from modernity, shows itself in an everyday symptom whose importance is often overlooked. When Moderns are put before a forest in free evolution – trees that are not exploited, are not cut down, are left in peace when they die and fall to the ground – they stop and deplore what they see as a state of abandonment or neglect. This modest observation reveals a metaphysics completely turned on its head. To believe that a milieu that is neither improved nor managed by humans is a milieu that has been abandoned, one must postulate that its necessary and intrinsic destiny is to be taken in hand by humans, whereas this space had existed for millions of years before humans came along (some mixed forests, for instance, are more than 200 million years old). To believe that an unimproved milieu has been abandoned is the soft madness of a paternalism that devalorizes the forces that have generated us.

## *Deconstructing ecopaternalism*

On the ground, on more than one occasion I have heard another even clearer formula, in the mouth of a hunter or a livestock farmer: "Without us, nature is a shambles." These few words express the absolutism of improvement: its subtle passage from the idea that improvement increases the usefulness of ecosystems for humans to the idea that improvement makes them better in themselves, makes them finally complete. This innocent little formula conveys the full violence directed by the modern dualist metaphysics against the nonhuman world: the outlook it expresses is an aberration that must be deconstructed before we can undertake diplomatic discussions with the practitioners of the earth – farmers, hunters, livestock breeders, tree planters, improvers. In short, we might respond: "The dynamics of life are billions of years old, they have made you; thanks for offering to help, but they get on very well without you."

The idea that "nature" needs, on principle, a hunter who regulates wild fauna, or a livestock farmer who shapes landscapes, reflects a paternalistic approach to the living world. This ecopaternalism is sometimes well intentioned, but that hardly matters: there is something absurd about being paternalistic toward the world that has made us.[8] Life, it must be said, is nothing if not paradoxical.

To put it simply, any discourse that asserts the need to control, manage, or protect nature "in itself" (and not for specific purposes, such as protecting the human community against famine) obscures the fact that nature first had to be viewed as dependent on humans in order to justify the need for humans to become its masters and attentive stewards. Contemporary critics of the modern relation to nature (perceived as despotic control) rarely see the camouflaged prerequisites of control. Control requires us to begin by heteronomizing the living world – that is, seeing it as something other than ourselves and thus considering it to be eco-ethologically deficient, dependent, erratic, unfulfilled; yet the very living world that we mean to guide and protect is in fact autonomous, in the sense that it has no need for us.

Ecopaternalism hides and represses the fact that it is *responsible* for the heteronomization, or "otherizing," of nature – that nature whose steward and manager it claims to be: it has invented out of whole cloth the idea of a "nature" presumed to be originally heteronomous and in need of our guidance ("without humans, nature is a shambles"). This occultation is an arrangement that gives ecopaternalism a clear conscience, allowing it to persuade itself that it is for the good of the Other that it is taking charge of its fate. We can call it "ecopatriarchy" when it transposes this mode of relating to all the alterities it holds to be inferior (women, foreigners, savages ... ).

The final operation in the process – *hiding* the original heteronomization of the living world – is fundamental. It is constantly replayed in descriptions of our various ethical relations to life. Philippe Descola makes this visible when he evokes protection as a relationship schema: "In relations with nonhumans, protection becomes a dominant schema when a group of plants and animals is perceived both as dependent on humans for its reproduction, nurturing, and survival and also as being so closely linked with them that it becomes an accepted and authentic component of the collective."[9] That a human collectivity considers it legitimate to protect its herds and its seedlings makes sense, but the intensity of the paradox must be highlighted: when that collectivity extends this conception of the living world that has emerged from a particular form of domestication to the living world *as a whole*, it attains the mild madness that has constructed our conception of the biosphere.

What living being is *spontaneously* "dependent on humans for its reproduction"? Not one. The very idea is absurd in evolutionary terms. For 4 billion years, the lineages of living beings have enacted their everyday prodigies, have interacted, constructed the habitability of the world, without our having had the slightest role to play in their maintenance, their survival, their prosperity – for evolution is precisely a force that continuously confers upon these lineages the greatest possible powers of prosperity and durability; that is the essence of evolution, since it regularly eliminates the variations that weaken the perpetuation of a species. In order for the necessity of protection to be established, the living world has to have been heteronomized – in reality, through control of reproduction, or in representations, through the invention of a metaphysics of the intrinsic defectiveness of "nature."

Let us imagine an agropastoral society that has invented a particular form of domestication (we shall look at a specific example further on), which implies making the domesticated life forms dependent, in such a way that this form of domestication plays an active role in their durability. But, in a second phase, this society proceeds to project onto the *entire* living world – the 10 million ageless wild species, older than we are, and the dynamics that produced them – the dependency that it has produced in the handful of species that it has domesticated. Hence the necessity of managing, guiding, and protecting this living world.

This is our history in a nutshell. Take an infinitesimal fraction of living beings. Make yourself master of their reproduction. Set up forms of relations that ensure their dependence on you. Transpose, fantasmatically, this status of deficiency/dependency to the entirety of the living world. Feel intimately responsible for protecting "nature." Now you are modern.[10]

To deconstruct this ecopaternalism, we need to set things to rights. To be sure, no, humans do not necessarily degrade "nature" by their presence, since they are living beings shaped by living milieus (exploitation is not condemnable in itself); but also, no, milieus do not *need* human activity to be valorized, fulfilled, organized: it is the human collectivities that need the gifts supplied by the prevailing ecosystems (exploitation is not amelioration in itself). This much posited, the questions become the following: What type of exploitation can be inserted into the living dynamics of specific milieus without shattering them? How can we modify these dynamics without mortgaging their resilience? How can we limit the damage done by exploitation to what is strictly necessary? We shall return to these questions at length further on.

Today, paradoxically, the heirs of the social sciences are the ones spurring this same confusion about improvement, especially in the light of recent eco-ethnological studies of the Amazon forest. It is important here to pay attention to a formula that is often uttered by those who defend milieus against any integral protection: they take the example of the Amazon rainforest to argue that their own autochthonous human uses, sustainable uses, have co-constructed the forest, and that therefore all integral protection of ecosystems should be banished, since nothing is "wild." In seeking to criticize the colonial fantasy of an intact nature, they nevertheless reintroduce the myth of improvement as an absolute principle, because they lack a crucial nuance. Reiterating that the Amazon rainforest has been modified by the indigenous peoples is certainly relevant to quelling the fantasy of the virgin forest and its primacy, but it is completely erroneous if it induces the belief that a forest is made *by* human users and fundamentally enriched by them. The Amazon rainforest is at least 55 million years old; Amerindians have been traversing and shaping it for, at most, 10,000 years. To postulate that the forest needs to be used by humans is to devalue its own timeless powers. We have to do justice to the real temporalities of this world if we want to escape fully from the modern anthropocentrism that overvalues our role. It is indefensible to bring it back surreptitiously on the wings of a clear anticolonial conscience. The Amazon rainforest, like any other forest, has never needed humans. Forests are autonomous milieus. That we can weave sustainable, sometimes mutualist, relations with them, as certain indigenous peoples have done, is a *fact* – but this will never mean that our action on a forest is necessary to its existence, its well-being, its vitality, or its durability; indeed, the opposite is true.

## Realigning Alliances

### *Restoring its full value to the living world*

From this starting point, an analysis of modernity by Raj Patel and Jason Moore can shed light on the conception of the living world that can be deduced from the idea of improvement when it becomes an absolute principle. In their indispensable book *A History of the World in Seven Cheap Things*, Patel and Moore characterize modernity as a series of processes for making nature "cheap."[11] "Cheap" means devalued, in all senses of the term: inexpensive, a good bargain, but also of low value, of little importance, interchangeable, and thus – this is my interpretation – not worthy of recognition or respect. This *History of the World* shows how the dominant modernity (the authors call it capitalism) has cheapened nature, work, people, care, and so on.

It is important to understand that the term "cheap," beneath its alluring familiarity, is an economic and ontological concept: the dominant form of modernity has "cheapened" the living world by labeling it "nature." In other words, it has transformed the living world into consumable matter of little economic value, but it has also devalued it ontologically, representing it as being of a secondary order of reality and importance, strictly a means to human ends: an inanimate décor, a reserve of resources. And the two devaluations go hand in hand – that is the heart of the problem. The ontological devaluation of nonhuman life was a strategy of economic depreciation: devaluing raw materials permits cheap production. This is particularly explicit with meat, a food product of very high ecological value, owing to the richness and exigency of the ecological processes that generate it. Its high value should have an impact on its costs and prices, should even inspire a form of reverence – but meat is devalued in our most recent economic forms by arrangements that conceal the negative externalities.[12] The dualism that reduces the living world to matter deprived of "interiority," of value, of meaning, is in fact a war machine serving to facilitate the extractive project of productivism ("it's just brute matter, we can do what we want with it").

## *The history of a double devaluation*

Adherents to the idea of improvement have been repeating for several centuries now that we need to enhance the value of milieus: forests, wet zones, wastelands. But the paradoxical aspect of this attitude becomes patent if we ask an indigenous hunter-gatherer of Southeast Asia or Amazonia, who views the forest as a giving environment that spontaneously showers on him all he needs for sustenance, whether the forest needs to be "made valuable." It is highly probable that he will not understand the question. For in our modern cosmology we have had first to cheapen wild milieus, through a covert philosophical act of denying their value, so that we could then believe in the necessity of "valorizing" them. We have had to cheapen the earth ontologically in order to legitimize the necessity of improving it and giving it *value*. The surreptitious displacement of the concept of value from its ontological sense to its economic sense (that of the surplus value of its yield) is what has allowed for this strange aspect of our inheritance.

This is the great, fascinating paradox of the whole business. In order for the theories of improvement to be imposed, it was first necessary to transform our conception of the earth by ontological cheapening – by stripping it of its value in the absolute. From this standpoint, nature on its own is deemed unfinished, still in its infancy, disorganized; humans thus need to intervene to "improve" it. But improvement, here, is once again a matter of increasing its yield, in

*Realigning Alliances*

terms of the biomass of grain and livestock needed to nourish human populations, which are increasing owing to the very fact of this surplus biomass. I am not questioning here the appropriateness or the necessity of ameliorating agriculture at the service of the effort to lift human collectivities out of famine, but I am challenging the philosophical underpinnings that have served to make absolute the idea that exploitation enhances value, and the kidnapping of the word "value" by the economic sphere. For, in a very short time, "improvement" stopped being a matter of allowing a field to produce more wheat to feed humans, and became a matter of producing more economic value to be injected into the capitalist machine; this is spectacularly visible in the contemporary paradoxes of the Amazonian deforestation. As seen by an Amerindian inhabitant, the giving environment that is the forest is in no way deficient in value; by contrast, it has no value for the forest exploiter or the Brazilian capitalist rancher, in the strict sense that they cannot draw direct profit from it.[13] The neocolonial conflict is a conflict of values bearing on the very nature of the milieus of life.

And this form of exploitation produces a second devaluation, a second cheapening, since the market logic applied to the gifts of the earth necessarily tends, if the negative external factors are ignored, toward the production of raw materials (lumber, for example) at the least possible cost: that is, cheaply. The portions of a milieu that are not put to work must be devalued (that is, considered worthless) in order to *justify* their economic exploitation conceived as attribution of value, and that exploitation then makes it possible to produce mass goods on an industrial scale: inexpensive, cheap, low-value objects. The living world is cheapened at both ends of the chain, and the first devaluation justifies the second. The first philosophical violence justifies a form of practical violence, which justifies a form of economic violence (the lumber, the meat, are not sold at their true ecological value), in the service of an overabundant consumption of cheap products. The aim of the entire arrangement, of course, is to generate wealth, with all the ambiguity this process entails: it is emancipatory in certain respects, in that it brings abundance to counter famine and lack, but it is alienating in all other respects as soon as this wealth is captured, confiscated, and made self-serving. The first devaluation is concealed to justify the second. I see this as a double devaluation of the living world by modernity.

This twofold movement of cheapening that carries us along, driven by the adherents of improvement, is constitutive of our recent modern uses of the earth. Progress appears self-evident, but its provincial oddness emerges when we compare it to the subsistence philosophy of indigenous peoples known as gatherers. Neither the devalued character of the original milieu nor the idea of improvement nor the reduction of subsistence to the production of cheap foodstuffs makes sense to them: the three ideas are equally absurd. In their world, there are giving environments that maintain themselves on their own and nourish humans with foodstuffs that are *precious by definition*, if we show adjusted consideration toward them. They do not see a deficient earth that has to be taken in hand in order finally to "enhance" it – that is, to put pressure on it and eventually destroy it. The course of history, on this point, is on their side.

## *Transmuting values*

This is why deconstructing this legacy of political economics is a central stake for a philosophy of life that poses the problem in terms of a transmutation of values: we must struggle against the insidious devaluation of the life within us and apart from us, a devaluation induced by the dominant modernity. Restoring to the living world its confiscated powers – communicational, interactional, semantic, ethopolitical, negotiating, world-making powers – is a way of conducting the cultural battle. For it is a battle that bears on the axiological apparatus incorporated in each of us – the apparatus we use to determine what has value and what does not; what is worthy of consideration, of respect, and what is not. Making visible, or inventing, the fact that the living world is the irreplaceable matrix of earthly existence, and therefore of human life, is a philosophical and political project, because it works against the hidden devaluation of the living world into cheap "nature" on the one hand and victimized "nature" on the other.

This formulation of the problem finally makes it possible, in the effort to identify the parties responsible for the ecological crisis, to see the enemies more clearly: they are all those who participate in the process of cheapening the web of life – that is, in the process that simultaneously devalues that web ontologically, depoliticizes it, and converts it into raw material for productivism. And the foundational act of modern naturalism has consisted precisely in giving a new meaning to the word "nature," a term that has played multiple roles from the Greeks onward. "Nature," in the dualist sense, is the name that has been given to the web of life, to our giving environment, now that that web has been devalued. Patel and Moore show this with exemplary clarity: "Nature is not a thing but a way of organizing – and cheapening – life."[14] In this light, then, it is ambiguous to assert, as a way of formulating the new alliances between social and environmental struggles, that "we are nature defending itself." Better to say that we are life that is defending itself – in part by combating its conversion into "nature."

## Rekindling Life

What late modernity has done in converting the living world into "nature" is thus not only an objectivization, a mechanization, a reduction to determinist causes, achieved by inventing the "sciences of nature" and their naturalist conception of the world. This philosophical transformation, on the order of ideas and representations, is the tree that hides the forest. What has to be brought to light is the *coupling* of epistemology and economics that has allowed the living world to be converted into cheap nature. This is not primarily an idealist problem of how to represent nature philosophically. The question, rather, is how images of nature induced by the sciences have been captured and diverted by economic and political machines (namely, capitalism, productivism, extractivism) to transform before our eyes the divine, sacred, giving Nature of the Ancients into the cheap nature of the Moderns. To combat that devaluation, the prodigy of life must be defended, its ontological coherence must be recalled. To bring the living world into the field of political attention is to free it from its recent, toxic status of cheap nature – to overturn its designation as worthless matter, as an inert, low-cost, available, accessible resource. To take this simply as a "philosophical" problem, on the order of pure representation, would reflect a latent idealism: it is in fact a problem of political economy, like all real problems. The challenge today is to invent the *antonym* of cheap, in order to redescribe life. We have to create philosophical, economic, political, emotional, and scientific arrangements for *overturning* the devaluation of the living world that we have inherited. We have to devise ways to fight that devaluation.

### *Trust in the dynamics of life*

What is the opposite of devaluation, or cheapening? The answer lies in consideration, gratitude, the search for adjusted consideration.

## Realigning Alliances

What is the opposite of the idea that the living world, without our action, is deficient, defective, incomplete? Trust in the dynamics of life. Trust in life as it has been unfolding in all its prodigal splendor on the surface of the earth for billions of years. Trust and consideration in the face of the dynamics of evolution and ecology that are weaving the web of life on a daily basis, that have made us as human animals, and that give us life every day, in the form of the oxygen provided by plants, and energy, and habitats.

We shall look more deeply, later on, into the implications of this inversion concerning the "productive" agricultural machine and the uses of the earth, but in passing we must stop to note that the inversion has made it easier to understand the profound signification of the defense of forests in free evolution. A milieu actively left in free evolution is a place that affirms and reminds us that life without us is not deficient. It does not need us. We can act on it, we can interact with it – we even have to do so, sometimes – but without ever forgetting that the environment does not suffer from a lack of our action, and that our action is not what ensures its vitality, its prodigality, its creativity: its action, rather, is what ensures ours.

This is, ultimately, the irreducible reason for defending the existence of spaces in free evolution. They are the standard bearers of a society that has put an end to its war against a nature deemed defective or hostile, a society that is finally acquiring a fundamental trust in the dynamics of life that keep us alive at every moment: a society that conceives of the old "nature" as its "giving environment."

The idea of a "giving environment," here, is a powerful concept for getting around the opposition between wild nature and nature put to work. The classic opposition holds that only nature put to work gives us something, the products of production, and that the rest of nature is there "for itself." This is the basis for the modern dichotomy pitting exploited nature against sanctuarized nature. The giving environment lives on a different map of the cosmos: for giving does not mean giving for us humans, it means giving for all forms of life, which in turn make life possible for all. This interdependence and mutual vulnerability are what keep us from posing the problem in terms of a hierarchy of interests between "nature" and "humans." Here is the magnificent paradox that establishes the identity of an ecosystem: the habitat of each living being is the web of all the other living beings; consequently, the web gives to all. And the word "giver" does not signify first of all "producer of marketable, exploitable goods"; it means that the giver generates the air we breathe, the water we drink, the landscape through which we traipse, all the non-appropriable dynamics of the living world that make life possible. Everything that is not marketable is *also* a gift, and what is exploitable in an instrumental relation is only a small fringe of what is given to all living beings. The challenge is to conceive forms for exploiting what is given by the milieu that do not weaken the majority of non-exploitable gifts: everything that is freely offered to all life and cannot be owned.

The dynamics of life are what make a milieu a giving environment. A nourishing earth, even in places not devoted to production – for even the dynamics that do not produce a consumable biomass shape the livability of the world. All marine algae help to trap carbon; every river makes life more viable. This is what ecologists call, clumsily, in the language of liberal ideology, the "services" rendered by nature. But these are in no sense services – the cosmos is not a service agency: they are gifts in the anthropological sense of the term, without purpose, and as precious as life itself; they call for adjusted consideration, trust, gratitude, reciprocity, and forms of counter-gifts.[15]

## Realigning Alliances

The fight to establish milieus in free evolution is thus a mechanism for reversing the devaluation of the living world that we have inherited. It is a rejection of the dominant modern image of "nature": that of absolutized "improvement," according to which human action is necessary to complete, ameliorate, and ultimately give value to wild nature.

Still, a hearth of free evolution is not a model or an ideal implying that all milieus ought by rights to become spaces of free evolution; that would be absurd, since we could not eat, and because the diversity of relations to living milieus – the diversity of uses of the earth – is a legacy that we must cherish. Nor are such hearths an exceptional remainder maintained to keep our consciences clear while we reserve the right to exploit the rest of our surroundings blindly, turning natural settings into museums free of humanity, rare examples of "intact" nature. It is important to move beyond these configurations that stem from dualism.

These hearths are spearheads, affirmations and activations of trust and consideration toward the dynamics of life. They are *among the forms* of adjusted consideration that we owe to those dynamics.

It is simply a matter of sowing throughout the landscape milieus free of the belief on our part that they need us: this is the essence of a hearth of free evolution.

Defending wild forests and rivers is thus not simply the opposite of taking improvement as an absolute principle: it is an antidote to that folly, an antidote designed to do its work more or less everywhere in the territory, as a reminder.

The hearths of free evolution are a way for us to learn *as a society* that the web of life is not deficient if we do not work it. It need not be "arranged" in order to be complete, and to "arrange" is not necessarily to "enhance": wet zones are not made more valuable when they are transformed into productive fields, except in the strict economic sense. Without this lesson before our eyes more or less everywhere in the French landscape, it is useless to undertake major agronomic reforms or economic revolutions, because it is the underlying software of the relation to the living world that is distorted.[16]

In a society like our own, without nature in free evolution – a society that no longer has free evolution because it does not trust the dynamics of life – hearths of free evolution are thus philosophically and politically urgent. It is in this sense that they have a privileged place on the spectrum of sustainable relations with the living world.

Ecologically, in the concrete sense, these places produce positive effects, but these alone will not save the world (if that messianic expression, inherited from the forced marriage between St. John and Hollywood, even has a meaning); philosophically, though, it might remind us of how things ought to be.

Once we have relearned that trust, as a civilization, once we have put it in place in the landscape, we can of course arrange our milieus, "enhance their value" economically, but *differently*, because the point of true north on the compass has changed: it is no longer a question of improving nature, but, in the light of trust in the dynamics of life, it is a question of gathering and inflecting those dynamics so as to allow the advent of a sober and sustainable prosperity.

## *No blind trust*

Appealing to a rediscovered "trust" in the dynamics of life does not mean that everything that is alive is desirable or perfect – that would be absurd. The living world is not "all love." Epidemics, destructive invasive species, all these also belong to the living world. But this cannot be the paradigm for what is: constraints and enemies cannot be the archetypal image of "nature." For the living world is above all our giving environment, and the one that has created us. There is no active, conscious benevolence toward us in the web of the living – no more than there is in a stream when you are thirsty, in a tree full of ripe fruit when you are starving.

And, of course, in the coming decades, the living world, disordered by our unsustainable activities, is going to unleash reactions that will give us a very hard time: crises in agricultural yields, the return of once-vanquished illnesses, epidemics. But the living world is not the cause of these crises; it does not have two faces – one, laughing and generous, that brings us wheat for bread, and another, pitiless and cruel, that brings epidemics and floods. There is only one world, and it is ours. It gives everything, and our challenge is to gather in the gifts that sustain our lives, and forearm ourselves as best we can against the gifts that do us harm, but without making war on the world for all that, since one does not make war against that which supports us.

Above all, finally, we must adopt no blind trust, no mystical complacency toward a resacralized "nature." The living world will give us neither norms nor models, it must not be the object of a cult, but there will be no alliance if we do not acknowledge its strengths and its autonomy, its non-deficiency. That acknowledgment is the basis for any real alliance.

It is with *full knowledge of what is* that trust is created: trust is not a subjective attachment (love of nature) or an inherited religious principle (respect for Creation). It is a lucid awareness of what we are – namely, living beings shaped and maintained in life by the dynamics of our living world. It is this full knowledge of what is that animates the indigenous peoples who live off the generosity of the forest; the scientific ecologists who are fascinated by the productivity in biomass of wet zones, as compared to that of human agrosystems;

and the agroecologist farmers of the Creuse region who observe daily that the soil and the bees are doing all the work, aware that they themselves are merely inflecting the dynamics.

A forest in free evolution is a witness to what life does when it is not put to work, and thus it bears negative witness to what active management does to life. It is a constant and reliable witness that allows us to evaluate the violence or the gentleness of the acts of exploitation we impose on forests, so we can correct our actions and enter into a relation of adjusted consideration. In the absence of forests in free evolution, we end up taking an exploited forest as a model for what a forest is: we end up believing that the forest needs to be exploited to be fully healthy.

A forest in free evolution is thus a witness, in the sense of a place that *attests* to the power of regeneration and reinvigoration of the dynamics of life, a place where we can regain trust in the world that has made us. But, as we have seen, it is also a refuge, if ecologic connectivity is ensured for a diversity of life forms. It is a source that radiates life throughout the damaged world that surrounds it. It is, finally, a standard bearer proclaiming trust in the dynamics that weave the web of life, our world.

A world without free evolution lives in distrust of the dynamics of life; it is a symptom of a society that has lost trust in its giving environment. To fight for that trust is to fight for the literal flowering of the places in free evolution that express trust without qualification.

## *Moving beyond the dualist opposition between sanctuarizing and exploiting*

Restored trust in the dynamics of life is something more than a concept allowing us to escape from the myth of improvement as an absolute principle and enabling us to understand the social and cultural function of free evolution. It is also the operator that will allow us to move beyond the dichotomy between exploiting and sanctuarizing. It is a compass we can use to conceptualize the real political alliances between the uses of the earth that the dichotomy contrasts artificially – for example, those of the Farmers' Confederation and those of associations that defend free evolution. It is the criterion for distinguishing not between exploiting and sanctuarizing but between two types of uses of the earth. The opposition between "trust" and "distrust" or "devaluation" allows us to assess different relations to the living world from the vantage point of a dividing line that is more serviceable in practice, and more liberating politically, in order to conceptualize the collective transformation of the uses of the earth that is necessary if we are to face up to the systemic ecological crisis. What we need is a redescription of the various practices on the basis of new conceptual approaches that can teleport us spontaneously outside the labyrinth of dualism – onto another underlying map.

If, as we have seen, exploitation is not condemnable in itself, the question comes down to distinguishing between unsustainable exploitation and sustainable uses of the earth. Of course. But where to pass the scalpel between them? Is sustainable exploitation simply exploitation that is a little more durable, a little "greener," a little better "compensated," a little less destructive than unsustainable exploitation? *Similar but less than?* No, for the model remains modern exploitation centered solely on yield and the interest of humans.

How can we understand and move beyond this model in the light of our new conceptual compass?

Let us observe the invariant traits of the dominant modern forms of exploitation, in order to capture them in a concept. I use the term "extractivist exploitation" to characterize the practices in agriculture or forestry that are industrial, intensive, often monocultural, and dependent on agrochemical industries. Sometimes called "conventional agriculture," which makes its hegemony visible, extractivist agriculture is a form of mechanized exploitation, partial to inputs,[17] and focused on large areas. Its watchwords are productivity and profitability. Like extractivist forestry, it entails a simplification of the milieus in question, in that it reduces the diversity of domestic plant varieties and wildlife. It is based on what is called the "optimal" (in reality, *massive*) use of chemical fertilizers, herbicides, fungicides, insecticides, growth regulators, and fossil fuels. Now, how are we to understand the philosophical meaning of this massive use of synthetic inputs? It actually constitutes a spectacular answer to the hidden ontological postulate of this approach: that the dynamics of life are inadequate for ensuring the profitability required of the setting. The function of a synthetic input is to substitute for the growth dynamics of the living world, deemed *defective*. From a philosophical standpoint, extractivist exploitation thus postulates the insufficiency of the milieu it exploits. Its fundamental attitude is that of *depreciation* vis-à-vis the dynamics of life: in other words, a relation of antagonism, devaluation, fragilization, and substitution. With this approach, it reveals its hidden philosophical identity: it is, quite simply, the agriculture that is heir to the ideology of *improvement*.

And the agriculture of improvement in effect claims a monopoly over the correct use of the earth. Now, the way it understands this correct use implies a yield of a sort that the dynamics of the living world in the vicinity *always fall short* of obtaining – because this agriculture is not adjusted to what a plot of land can produce, but rather to the above-ground mechanics of the markets that fix prices in an economy that is not articulated with real ecology, the dynamics of life always appear *insufficient*, and call for substitutions, short-term forcings that are destructive in the long term (heavy applications of fertilizers and pesticides, violent treatment of the soil).

It is not that these farmers are incompetent or immoral; it is that the system of prices has repercussions on their actions to the point of producing, through feedback loops, the insufficiency that is virtually constitutive of their lands and the dynamics of life that shelter them: it is a crazy-making arrangement, as is evidenced by the tragic tally of suicides among the farmers who have inherited this tradition.[18]

The characteristic feature of this use of land is, finally – as we are recognizing fearfully today –that it is literally unsustainable: that is, it weakens the very conditions of possibility of its own perpetuation, by killing the life of the soils, destroying the populations of insects that are its pollinators – to the point that we have to imagine, in an ultimate form of substitution, using drones as pollinators to replace the bees, insects, and birds that it has destroyed by its massive use of pesticides.

Rather than change practices in order to allow the populations of pollinators to regenerate on their own, extractive agriculture, in its Promethean pride and its devaluation of the spontaneous dynamics of the living world, prefers at present to replace them with underpaid workers, dedicated to pollinating each flower with cigarette butts imbued with pollens, attached to selfie sticks.[19] The image of this quite real practice of artificial pollination is the clearest symbol I have found of the philosophical essence of this agriculture: it is based on *distrust* and *depreciation* with regard to the dynamics of life that underlie it, and thus requires human labor to substitute for those dynamics – this labor, too, is cheapened in a rebound effect. One depreciation implies the other.

Thinking in terms of trust in the dynamics of life, as opposed to distrust (requiring *improvement*), thus gives us a new way to grasp the implicit philosophy behind intensive agriculture based on inputs. That agriculture is defined by a drive to substitute human power for the powers of the ecological dynamics, rather than activating those powers and taking care of them; it is defined by the will to operate by imposing force on a milieu, rather than by carrying out the delicate act of inserting oneself into its dynamics to inflect them (this latter being characteristic, as we shall see, of intensely ecological farming agroecologies).

## *Portrayal of pests as "wild nature"*

Given that its model is constructed by exogenous imperatives on the order of the market economy, which is blind to the living ecology of a milieu, this unsustainable agriculture, dominant today in the so-called developed countries, ignores, obscures, and denies its real relation to the dynamics of life. This blindness is perfectly visible in a symptom that can be detected on the ground: the relation that the heirs of these forms of exploitation maintain with what they view as "wild nature," and with what they view as "pests."

There is in fact a tendency in this inheritance, encountered again and again on the ground, that consists in conceiving of wild nature as an external force that intervenes to parasitize production. But this conception is a deleterious effect of a dualist philosophy. Photosynthesis is wild, and this is what gives us each grain of wheat; the millions of years of coevolution between hymenopterans and flowering plants are wild, and this is what makes fields work; pollination is wild, and this is what brings back the harvest every year. So when a wolf turns up, or an ash aphid that devours apple trees, or a parasitic codling moth, or a buzzard, it's called both a "pest" and "wild nature" in a way that sets up the former as the official representative of the latter, and this entity – the pest – is deemed misplaced; wild nature is seen as *opposing* the processes of agriculture. It is like hating one's entire body and treating it as an obstacle because one has caught a cold. But the body is the prodigy that permits *all* of life – including, sometimes, colds.

Setting up the "pest" (which never exists as such in itself, but only in relation to particular interests) as the archetype of wild nature is a philosophical bias of this extractivist agriculture of improvement, which generates intolerance toward wild biodiversity, seeing it as nonproductive and calling for its eradication.[20] This is, moreover, what distinguishes it from agroecologies, which maintain much more ambivalent, nuanced relations of controlled hospitality toward the bioaggressors that threaten its crops; these relations of biological struggle turn the dynamics of predation against proliferation, a dynamics of vigorous but never eradicatory negotiation (what is at stake is the maintenance of proliferation below a threshold of viability for exploitation).

## *The wild forces that make a farm*

To make this argument clear, I need to introduce another actor – one that will represent the discourse and the cause of agroecologies: that of organic farming, which is richly endowed with a different relation to its milieu. Just a few kilometers away, as the crow flies, from the hearth of free evolution recently acquired by ASPAS (the Vercors Vie Sauvage reserve), we find the Grand Laval farm. It consists of 14 hectares (about 35 acres) of polyculture and stockbreeding, and orchards that serve as pasture to sheep and chickens: a whole system very carefully thought out so it can become autonomous with respect to inputs, and capable of feeding humans even as it invigorates the milieu.

*Rekindling Life*

The Grand Laval farm can be understood as a significant lever of ecological action, one of the scalable arrangements we need to rekindle the embers of life. A lever of this sort always consists in a local and circumstantial encounter, on the ground, of several seemingly discrete ideas that are powerfully transformational when taken together, without fanfare. The Grand Laval lever consists in three little ideas that the head of the farm, Sébastien Blache, came up with: a circular polyculture combined with animal raising, all organic; a short-circuit marketing system allowed by the creation of a producers' store; and a form of proactive hospitality that permits the greatest possible wild diversity. These three practices work together to protect the farm.

If one spends some time pruning pear trees and running after chickens with Sébastien, one can experience a significant metaphysical phenomenon that is often neglected – even if all farmers who are not blinkered by agricultural modernity have always been aware of it. The terms "domestic," "agricultural," "exploited" basically just refer to the timeless wild dynamics of the living world, but inflected *around the edges* by human activity.

What are these dynamics that weave the web of life? Photosynthesis, pollination, facilitation, cooperation, predation, parasitism, mutualisms. In more structural terms: variation, selection, diversification, and thus coevolutions, regeneration, resilience. From a still more global ecological standpoint, we are talking about carbon cycles, raw materials, water, flows of solar energy streaming into the trophic pyramid.

All this is activated when someone cultivates wheat or apples, when someone harvests wood for construction. Are these dynamics – let us come back to the question – wild or domestic? We can see now that the question is badly formulated: as a well-known Amerindian proverb puts it, before the colonizers arrived with their idea of domestication, nothing was wild. The dynamics at issue are wild in the sense that they are life as it was, long before we came along: they are independent of us. We did not create them; they created us. But their wild character does not exclude them, as the exclusive dualist logic would have it, from places where the living world is domesticated. Because what is domestic is just a way of bringing the dynamics together and inflecting them.

As soon as we think in terms of the concepts of the dynamics of life provided by the sciences of eco-evolution, the dualist opposition between wild and domestic looks like a misguided mindset: in fact, there are only wild dynamics, some of which are locally and superficially inflected to our advantage by artificial selection and human labor, sometimes disfigured and weakened in the process. Nevertheless, the totality of their functioning and their effectiveness is far older than we are.

## *Metaphysics of production*

All this sheds new light on a seemingly anodyne statement, repeated in every manual of universal history, on all the informational signage in museums of prehistory: with the Neolithic emergence of livestock raising and agriculture, we are said to have passed from "predation to production."[21] This claim constitutes our new enigma, and it serves as the first trace of a new path to follow in this inquiry into our relations with the living world. The canonical formula by Vere Gordon Childe has become our founding myth. The birth of states, of cities, of sedentary living, is said to have come from agropastoralism conceived as *production of our subsistence*, as distinct from the practices of the hunter-gatherers who were purportedly content with predation applied to the natural "reproduction" of animals and plants.[22] Let us not linger over the residues of imperialist ethnocentrism in such a distinction; let us rather focus on the very meaning of the intriguing formula: passing from predation to production, thus producing one's own subsistence. As soon as we analyze this statement, its self-evidence disappears and an abyss opens up.

We were hunting mouflons, and now we eat sheep. Etymologically speaking, "to produce" originally meant "to give birth."[23] But it is not the livestock breeder who brings the fully equipped animal from his own flesh, but the ewe who gives birth to the lamb, like the mouflon ewe who gave birth to her offspring. And isn't it the sun that makes prairies grow, and prairies that make both mouflons and sheep grow? But then, in what sense can we say that we "produce" the ovines that were living in Europe even before *Homo sapiens*, or the wheat that has existed for several million years?

What can this really mean, if we look closely? How could such a mysterious metaphysical operation have been established as a ready-made self-evident formula to define what we have become? We are really confronting a mental mutation, a blind spot so powerful that even the major theorists of the Neolithic continue to repeat it unquestioningly. When twentieth-century archaeologists speak of "production," they are projecting on the remote past a concept that they have inherited from their modern theoretical tradition, projecting economic, philosophical, and theological prejudices that have grown out of the Western tradition – and that need to be deconstructed.

Posing the problem no longer in the dualist terms of domestic fields versus wild milieus but in terms of the dynamics of life gives us a new way of looking at this fascinating history.

To be sure, a farmer can clear the land and replace a wild forest with a wheatfield, but still, when he cultivates wheat, he is not *producing* it, in that he is not an author fabricating it from passive material.

What does "producing" mean, then? Philippe Descola, when he set out to determine the mode of relation so natural to naturalism, known as "production," defined it this way:

> As a way of conceiving action on the world and a specific relationship in which a subject generates an object, production thus does not have a universal applicability. It presupposes the existence of a clearly individualized agent who projects his interiority on to indeterminate matter in order to give form to it and thus bring into existence an entity for which he alone is responsible and that he can then appropriate for his own use or exchange for other realities of the same type.[24]

In this sense, no farmer has ever *produced* wheat or mutton. What produces wheat is not the "projection of [a human] interiority" onto "indeterminate matter": the prodigiously complex organic form of wheat comes from millions of years of evolutionary shaping. The domesticator of the past merely intervened every year to select the wheat that bore the biggest grains and the ears endowed with spines that did not break, two modifications among the *spontaneous* varieties of wild wheat. Even in the visible transformations between wild and domestic wheat generated by domestication, *evolution* was doing the work, producing both the non-directed varieties and the selected one (with the farmer adopting the mask of selector for the occasion). Darwin came across this truth by observing the work of growers, noting that man "can neither originate varieties, nor prevent their occurrence; he can only preserve and accumulate such as do occur."[25] Darwin uses the word "originate" to deny paternity to the producer of races – to preserve or favor certain spontaneous propositions of the living world is not the same thing as to originate or to produce.

*Rekindling Life*

What *produces* wheat, then, is photosynthesis modulated by the long evolution of cereals: that is, their irreplaceable capacity to take nourishment from solar energy and use it to create living matter in the form of grains. Wild wheat is the opposite of "indeterminate matter": it is a highly functional form endowed with the power to metabolize inorganic elements, which no human technology can reproduce.

As for what produces mutton, once again it is photosynthesis and the evolution of ruminants – predating our own – that have invented this metabolic prodigy of generating flesh via the digestion of plants in the rumen, enabled by a symbiosis with bacteria. No human produces meat; we are destined to profit from this timeless multispecies prodigy.[26] In the pastoral tradition that concerns us, the idea of "animal productions" rests on philosophical operations that serve to remove any possibility of considering that animals themselves work or produce value.[27]

What do we do, then, if we do not *produce* (for there is indeed work – demanding, harassing, intelligent labor – in farming)? We *gather* the timeless powers that come from the coevolution of a lineage with its milieu, and we *inflect* them in order to improve the diversity of harvests, to increase their abundance, their flavor, their durability.

And yet, in the modern legacy, all bureaus of agriculture and economic systems speak of "agricultural productions." Farmers have become producers. How have we been able to base the history of what founds our mode of subsistence on a formula that is belied by the slightest observation of a wheat field? How can we consider that we produce what nourishes us? This ontological cataclysm is a great mystery.[28] Myriad agricultural and pastoral cultures around the world have domesticated living beings without ever saying or thinking that they "produce" them.[29] In other words, the idea that humans "produce" is not a necessary consequence of the passage to agropastoralism – it is a founding myth. To understand this myth fully would require a long effort of excavation, but I should like to say just a few words about it here. This is not to write its history, but to clarify its consequences through the interplay of multiple comparisons with other relationships to the living world: in short, to sketch out a morphology.

The fantasmatic idea according to which we *produce* our sustenance is what I call the "metaphysics of production." This metaphysics can be described as a philosophical, economic, political, and legal construction characterized by the addition of original properties to the human activity that will be henceforth called "production." How does it work? It follows a hidden protocol that can be traced. First, it has to devalue the agency of life in its own genesis – that is, it must minimize the role of the ecological and evolutionary forces necessary for making wool and meat and grain. Next, it has to overvalue human initiative in the genesis of the "product" (the transformation costs, so to speak). Through this gesture, producers set themselves up as "authors" of the living entities that they cultivate or raise, thus spontaneously allowing themselves to appropriate these entities (this operation is a condition of possibility for making such entities private property).

A highly telling example of the overvaluation of human initiative in the genesis of a food product that is to be positioned as "production" has been analyzed by Michael Wise in his history of the relations among capitalist ranchers, livestock, Blackfeet Amerindians, wolves, and bisons on the frontier between Montana and Alberta in the nineteenth century.[30] Wise shows that the categories of producer and predator are not ecological realities; rather, they are contrastive representations *constructed* by the colonizers in order to legitimize their land grabs and to sanctify their struggle against the former inhabitants of the milieu, humans and nonhumans alike. In fact, to justify their status as *producers* of beef, as creators of value (and also to differentiate themselves from the Amerindians and the wolves, which are, according to them, only *predators* of bison, and thus destroyers of value), the capitalist ranchers have to take on the role of *protectors* of livestock – even though they are raising the animals ultimately to kill and eat them (acting as predators, then, in the ecological sense). Protectors against whom? Against predators. It is here, by protecting one's prey against predators, that one can set oneself up as a "producer." That is all it takes – even though, from the standpoint of functional ecology, there is no difference between the energy circuit that goes from the sun to the grass to the bison and then to indigenous persons (the latter stigmatized as predator) and the circuit that goes from the sun to the grass to the cattle and then to ranchers (elevated as *producers*, creators of value). The same actual ecology, a different metaphysics. A producer, in this pastoral tradition, is thus a predator who spends his time protecting animals against *other* predators; whereas those others (Blackfoot or wolf), for their part, pass their time as predators in search of their prey. This justifies the producer's demand that the others be either educated or eradicated. It is from the angle of the difference in everyday practice between two meat-eaters that the essential difference in this strange metaphysics is established, along with the ensuing politics.

By this metaphysical power play (devaluing the agency of living entities in their own genesis, overvaluing human initiative), this tradition has appropriated domestic plants and animals for itself, by coding them as *products* of our action. The decisive effect of this process comes down to extracting these living entities from the model of exchange with a "giving environment" that is proper to animism and to hunter-gatherers' modes of subsistence,[31] and considering them henceforth as "indeterminate matter" to which our action is supposed to have *given a form* of which we are supposed to have been the *authors* – thus, the owners. Here we see a logico-political link between production and appropriation. The myth according to which we produce our own subsistence is necessary in order to justify the appropriation of living entities. This link between production and appropriation appears transparently in the contemporary question of whether a living entity can be patented, a question that *restages* the metaphysics of production. The industrials involved with genetic engineering are content to modify *one* gene among hundreds of thousands, a trivial fraction of the genome of a living being, so as to take possession of the variant thus constituted, and to appropriate for themselves a species that has in fact evolved over millions of years.[32] By analogy, walk up to a famous painting in a museum – a work by Tintoretto, let's say – and add a touch of color in a corner with a felt marker, and it's yours, it's you who've made it, because you've "produced" it. Here we can see the touch of magic in the metaphysics of production, which converts an ontological fantasy into a political economy.

One might object that the varietal selection that takes place in laboratories in modern agronomies, along with transgenesis, genetic engineering, and mechanization, all accentuate the role of human genius in the productive process. But in fact, if we take the long view, these technologies are never anything but a superficial inflection with respect to the actual productive forces at work – that is, the dynamics of ecology and evolution. The technologies increase yields, but they do not invent *any part of* the process.[33]

Given all this, the major ecological effect of this foundational myth of our society is that it cancels out the timeless motif of a *debt* toward the nonhuman milieu, a motif omnipresent in the animist cultures that do not believe in the fable according to which they produce the manioc they cultivate. What this cancellation allows, consequently, is a deregulation of *limits* in the exploitation of the milieu. It is precisely because animist hunter-gatherers acknowledge their debt to the forest for their very survival that they recognize the forest as a giving environment, and thus that their social form can spontaneously impose *limits* on everyone in the use of the commons – that is, on the products of the dynamics of the living world. Recognition of the debt induces very powerful social mechanisms that establish forms of durability.[34]

From there, in a last decisive effect on our agricultural and social forms, the myth of production cancels out the need for *reciprocity* in relation to the giving environment. Since we have produced our own substance, rather than having received it from our milieu, we owe *nothing* to *anyone*. This is a challenge to the motif, widespread among human societies that do not believe that they "produce," of gifts and counter-gifts in relation to nonhuman entities: the motif of the need to *restore* something to the milieu that keeps us alive – the need to maintain sufficient vitality in that milieu for it to regenerate and keep on giving. In the metaphysics of production, any gratitude toward the non-intentional gifts of the milieu is inconceivable.

To represent the change in worlds implied by the metaphysics of production schematically, we can sketch out a set of morphological contrasts, a fictional tableau comparing realities that come from heterogeneous histories and geographies. In the subsistence mode of the hunter-gatherer, the living entities that provide sustenance are beings, bearers of flesh whose life we negotiate with an intercessor, the gamekeeper or the master gardener, in a conception of the world viewed as circulation of flesh among different forms of life (humans, animals, plants, forests, tutelary powers): a horizontal circulation that requires exchange and reciprocity between humans and non-humans.

The herders and farmers who see themselves as producers live in a different cosmos. Here, animals and plants are no longer persons whose gifts are negotiated with the master gamekeeper or gardener, but goods that these "producers" are free to dispense: they have substituted themselves functionally for their ancient gods. The herder controlling the reproduction of domestic game takes the place of the old gamekeeper – thus, of the one who dispenses. It is the herder who chooses what animal will be killed and shared by the group (this is the motif of *méchoui* in pastoral cultures, where this transcendent status is conferred on the patriarch). Dispensers are exempt from relations of exchange with the living beings that they dispense: they feel fully entitled to dispose of these beings as they wish. This is the first time that the human animal resembles a god so closely.[35] There is no more need to negotiate the circulation of flesh among species, because dispensers are the proto-owners of these forms of life in their enclosures. They no longer owe anything to anyone, not to the gamekeeper, not to the milieu: the circulation of flesh is shifted into the closed circle of a *single* species, *Homo sapiens*, in the form of trade and inheritance; and animals, no longer viewed as persons, become appropriated goods. (To be sure, this mutation took place over several thousand years, but we have embarked here on a pedagogical fiction.)

The animals in question have become wealth, objects of exchange among humans; they are no longer flesh-bearing persons, subjects of exchange among life forms in the giving environment.

As the founding myth of the Moderns, the metaphysics of production thus operates an essentially political gesture that authorizes two things: the affirmation that humans *produce* permits the appropriation of living entities, and it justifies shedding any requirement of reciprocity with respect to the giving environment. This is the function of the myth, and its mechanism for eliminating limits. Turning the gifts offered by the giving environment into productions is in a sense the foundational act of naturalism.[36]

Simply put, the idea of production is an original and quite strange conception of human action, even though it has been naturalized in our tradition. Production is characterized as a particular superpower: that of bringing what I call a-nature into being on the basis of nature. Production thus invents *a-nature*, deficient nature (the *a* is privative); the "other" of "nature" then comes to qualify everything that stems from culture, technology, and the social realm. This view of production is an undisclosed condition of possibility of the familiar dualisms: nature/culture, nature/humanity, nature/artifice. By this means, the heirs of this operation conceive of themselves as beings that have extracted themselves from an encompassing whole (the milieu of life), which, by contrast, comes to be called "nature." "Nature," in the sense of the world not produced by humans, is what is left over, what emerges and stands apart as external, once we conceive of humans as living among *their own* productions – like this fragment of nature whose action has its *own* magic (that of generating a-nature). The Moderns' "nature" is a residue.

In naturalism, the fact of having interiority is thus not really what marks the distinction between humans and "nature" – it is, rather, a *symptom* of the possession of the originality that establishes the real distinction, which is hidden. This originality is the fact of possessing a unique power, attributed to humans alone: intentional action – and, more precisely, a *certain conception* of human action, which is defined here as an event that *produces* things and is not determined by external causes. The possession of interiority is thus only a secondary symptom of what ensures the capacity of humans to *escape* "nature." The naturalist standpoint does not maintain that everything human, or the human in general, escapes nature (the physical body indeed belongs to nature, as does the neurobiology that supports symbolic life, and the matter of which every technological object is made) – rather, what distinguishes humanity is a *supernatural* capacity, imagined by our tradition: intentional action as the capacity to produce. We can see here why "nature" is actually a bizarre ethnocentric concept. By contrast, "a-nature" is not what is human in general, but human action and its products (thus, cultural creations, technological inventions, social institutions, and so on).

Our exotic metaphysics concerning human action thus claims to have invented a lever of transcendence, an alchemical transmutation, from nature itself. According to this metaphysics, there is a moment when, from nature, something other than nature emerges, and human action is precisely the operator that creates a-nature: something of a different nature from nature, something that escapes nature even though nature is its source.[37] What a strange philosophical legacy we have – it can hold its own against the most extravagant mythologies!

This self-realizing metaphysics has emancipated humans under the sway of its legacy from their constitutive relations of gift and exchange with the giving environment; thus, humans are released, freed from their interdependencies. Through the belief that human action makes humans the authors and owners of the ontological clay of which they have taken hold, humans effectively remove themselves, offering themselves an ungrounded and potentially destructive freedom, like the cosmic solitude that also characterizes the Moderns. Using this power to create a-nature in order to transcend "nature," to escape and supplant it by their actions alone, becomes one of the missions of these strange humans.

## Rekindling Life

All other peoples certainly create technological objects, act on and develop, in part, what they consume, but only the heirs to the metaphysics of production call all that "producing." There are in fact numerous ways to conceive of human action that do not involve thinking in terms of production: human beings are vectors, the sites of passage of greater forces that traverse them and that they inflect, with which they engage in exchange or negotiation. But we are heirs to a magical conception of action (most of us, anyway, even though this conception is challenged by myriad theoretical and practical counter-attacks), in which we are not vessels of forces that exceed us and irrigate us, but uncaused causes, absolute departures, and bearers of internal forms that we impose on passive matter. And, consequently, we are producers.[38]

The metaphysics of production can be envisioned as a massive machine operating today in a unified way, although multiple forms of alternative relations to action, to the living world, to the earth, are manifest all around it; these alternatives have endured, but in limited and marginal forms. To put it differently, the discreet architectonic metaphysics of production is probably hegemonic in the West, even if it takes several different guises; that metaphysics remains everywhere, in us, among us, on our territories, in our practices, our relationships, and our ideas.

To synthesize the argument, I can list the various dimensions of the metaphysics of production: these have multiple origins, but they have crystallized in modernity.[39] First, this metaphysics is an original philosophical conception of action (authoriality, hylomorphism, appropriation, devaluation of what is received and overvaluation of the costs of transformation).[40] It is also an anthropological distinction (only humans produce; the other animals, the milieus, thus cannot generate anything of value). In addition, it is the hidden operator of our original dualist opposition between nature and a-nature (culture, society, artifice). It is also a theory of the "proper use of the earth" that ends up essentially serving as the justification for European land grabs (it is legitimate to seize the lands of peoples who do not produce, since those peoples are failing to fulfill the destiny of the earth). Finally, it entails a certain relation to the living milieu: that of improvement as an absolute principle, as we have already seen.

## *The difference that makes the difference*

Armed with these concepts, let us return to our farmers. Every agricultural exploitation, as we know, functions as a whole only by relying on the timeless dynamics of the living world and receiving its gifts: it does not *produce* those gifts, does not offer substitutes for them; it enjoys them as coming from a giving environment whose expression it can modulate – or which it can destroy.

This is an important preface to a discussion of the relations between agriculture and wildness. And farmers who come from time-honored agroecological traditions are very often aware of this: the dualism implied by the principle of improvement is not reflected in their practices or their outlook. I remember a remark made by a woman from the Drôme Farmers' Confederation, a sheep breeder whose thinking went serenely beyond the metaphysics of production: "I am not the one who produces wool; photosynthesis does that."[41]

The only important difference, then, does not lie between the foodstuffs we produce (harvests) and those that we do not produce (the wood from wild forests, for example), since we actually produce none of these, strictly speaking; the difference lies between exploitations that mortgage the giving environments, wring them dry, weaken them, and prevent them from cohabiting with wild biodiversity (the biodiversity that is not productive in the economic sense), on the one hand, and exploitations that merge with the wild giving dynamics, cohabit with them, and maintain them, on the other.

If we formulate the problem in terms of the dynamics of the living world, the opposition between exploiting domestic nature and sanctuarizing intact nature becomes meaningless. The problem becomes one of favoring the dynamics of the living world along a whole gradient of milieus; it is a matter of rekindling the embers of life, in both exploited and unexploited milieus, and acting against *all* the practices that weaken and devitalize these dynamics.

From this point on, the absurdity of the forms of agriculture and forest management that are based on antagonism toward the timeless wild dynamics of the living world becomes eminently clear: these latter constitute everything that keeps an exploitation alive, which means that conventional agriculture is based on a destructive defiance toward its very foundation – toward all the forces that make it work.

And we can now understand better, by contrast, what characterizes agroecology. I shall use this term from here on to refer to any practice that intends to "make do with," in every sense and in the strong sense – "with," and not "against" or "without" – the spontaneous dynamics of the living world on a farm or in a forest according to their own requirements. In other words, agroecology does not wager on the domestic against the wild; instead, it bypasses the distinction. The dynamics that constitute agricultural production in its entirety have been the driving forces behind ecosystems since long before humans entered the picture.[42] This is an agriculture that works peacefully *outside* of the metaphysics of production.

This approach is clearly visible in the practices implemented on the Grand Laval farm, for example. Indeed, it lies behind the farm's most obvious originality: owing to his past as an ornithologist, committed to the protection of birds as a salaried employee of the League for Bird Protection, Sébastien Blache has taken the practice of hospitality toward wild avian life to a high level on his farm. In his orchards and sheep pens, one finds a plethora of nesting boxes for tits and raptors, as well as roosts for bats. In addition, there are arrangements offering hospitality to wildlife in general: ponds and woodpiles, uncultivated areas – all these add up to "meals and lodging" intended to bring wild creatures back in.

What might this approach signify? Sébastien Blache's reasoning extends beyond privileging auxiliaries, the form of biodiversity known as "functional" – the form that is understood to put itself spontaneously at the service of exploitation, by devouring creatures harmful to harvests, for example. Blache makes room for a large number of species, without prejudgment as to whether or not they are useful in one way or another. Above all, he lets them come back in *sufficient densities*. This is the heart of the matter, as he sees it, and it reveals, if we look closely enough, what interests him in this business: focusing on the density of tits rather than on bringing back the largest number of *species* of tits, independently of their population size, amounts to posing the problem in terms of *dynamics* – density is what is required for these species to *play the ecological roles* that characterize them. For it is starting from a certain population density that tits will be effective predators against insect larvae when it is time to feed their young. What motivates Blache is not simply an aesthetic interest in birds, or a desire to check off as many species as possible, or a moral sense that they have a right to exist; his main purpose is to allow the timeless dynamics of life to reconstitute themselves on his farm. Bringing back as much wildlife as possible does not consist in instrumentalizing certain species in the work of the farm, but in reconstituting the dynamics of an ecosystem as best he can. Owing to his scientific studies and farming experience, Blache knows that these dynamics confer resilience, sturdiness, vitality, and prosperity to the entire milieu; in other words, he practices a kind of well-informed trust. He knows that these dynamics protect the web of life, and thus also, within that web, his own farm.

In the words of the exploiter himself, the agroecology practiced on the Grand Laval farm manifests, far from the distrust requiring *improvement*, a trust in the dynamics of life.

## *The race to disarm*

How are we to understand with precision the idea embodied in the Grand Laval farm: that the wild dynamics of the living world *protect* the milieu and, at its heart, the farm itself? This is a central aspect of trust – one that is harder to grasp at first glance, however, than recognizing the living world as a giving environment, for example. To understand it better, we need to make a detour through a geopolitical analysis of the ecological relations among species.

Let us adopt the viewpoint of a species of bioaggressor – a "pest." Let me imagine myself for a moment as an insect arriving in a rich milieu. As soon as I proliferate, I become a resource for others, I open up an ecological niche for those who might use me as food; I create an opportunity that may be seized by another life form to adapt to me as an abundant prey, or as a parasitable host. Here is the full beauty of evolution applied to ecology. Every proliferation in an ecosystem implies regulation, *as soon as* there is enough biological diversity in the vicinity: the diversity of species, populations, and genomes is the condition, here, for the adaptive potential of the milieu when it is a matter of limiting the damage caused by the proliferating species – the conditions, thus, of resilience.

## Realigning Alliances

We sometimes observe, in an ecosystem, a situation that resembles the "law of the jungle," or a war of extinction: the unlimited proliferation of a species. The most spectacular example of this phenomenon is probably the brutal biological invasion of an ecosystem by a newcomer, where the absence of coevolution over time has led to the destruction of its competitors for the same ecological niche, and of its prey. But this phenomenon is not the ecological norm; it is, rather, an exception, since by definition it cannot *last*: once time has done its work, the proliferating species is limited on all sides by the evolutions of its competitors, of the predators that feast on it, of the parasites that profit from its increased density. We can deduce from this that the immense majority of ecological relations that we observe in the ecosystems around us reflect a stabilized cohabitation in which *no population* retains the power to proliferate to the *absolute* detriment of another – for coevolutions have invented and reinforced the conditions of a modus vivendi (defenses, adaptations, predations, mutualisms, and so on). The evolution of an ecosystem, by constantly generating diversification and the mutual adjustment of life forms to one another, has the effect of limiting proliferations (this is why the world is not completely blanketed with Darwin's orchids). Needless to say, we cannot pinpoint all the myriad interweavings that produce this effect, but we observe it nevertheless.

To grasp the phenomenon of coevolution, biologists have used the metaphor of an "arms race": a lineage of prey will evolve adaptations in order to struggle against its predator, which will evolve counter-adaptations to keep on hunting that prey.[43] This metaphor is applicable in certain specific contexts in evolutionary biology. But beyond its unnecessarily military connotations, it is actually problematic when it comes to understanding the *ecological sense* of coevolution, because it focuses on just *one* relation between two species, isolated in abstraction from the rest of the ecosystem. In fact, if we take into account the *multiplicity* of the relations within an ecosystem, coevolutions constitute instead a *disarmament race*, leading toward a balance of power. It is a struggle for the *preservation of political relations* between species. By "political," I mean a space of relations that prevents full-scale war of all against all. Plural coevolution in fact always ends up preventing one single invasive or destructive species from "refusing" political relations: the other species arm themselves to make a given pest dependent and limitable, force it to cohabit, to make concessions, to reposition itself in the game of mutual vulnerability we call an ecosystem. This is, of course, not an intention or a project, it is an invention of the vital play of a milieu, induced simply by three ecological properties of the living world: it is an intrinsically plural world; each member of its plurality seeks to live and deploy itself; each member nevertheless remains dependent on many other such members. The balance of power here is analogous in certain respects to the balance of power between nation-states as characterized by the realist theory of international relations. However, its direction is reversed, because interdependency is constitutive of the forms of life present in any given milieu, whereas nation-states define themselves in exclusive terms. The territorial logic is also reversed: the existence of exclusive sovereign portions of space is an anomaly, not the norm (for any species with which I enter into a relation of competitive exclusion in a space, there are a thousand others nearby with which I cohabit – thus, the logic of national territories with fixed borders makes no sense here). In the living world, the other members constitute my territory of existence; we all slide in and out among one another's lives. Imagine a multilateral geopolitical situation on an interwoven shared territory, a space of everyday political cohabitation. This phenomenon hybridizes certain properties of what the Moderns have radically defined as distinctive differences between interstate geopolitics, on the one hand, and politics among individuals cohabiting on the same territory, on the other. Thus, it cannot be analyzed through the traditional categories of political science.

The race to maintain political relations amounts, in fact, to an arrangement of cohabitation through interweaving. This does not exclude conflict; rather, it incorporates conflict into a more ample and more ambivalent problematics. It is an adjustment that struggles to reduce the unlimited character of a power that has been operating to the detriment of those with which it is interwoven, by binding it so closely with those others that it can no longer go it alone in any respect. It is thus an arrangement that leads an interdependent entity to remain within the space of vital negotiation, preventing it from exiting that space, not by mutilating or intimidating it (as in human geopolitics), but by tying its destiny imperceptibly to that of all the others. Reweaving the community in place, exacerbating the mutual vulnerability of its members – this is the unintentional strategy invented by the living world to solve the politically impossible equation, always to be returned to the loom: living in common in a world of alterities.

This space of negotiation is what I have called elsewhere an ethopolitics.[44] It consists in tinkerings with the various modi vivendi that are interwoven among interdependent life forms. These interactions are indeterminate, historical, changing, fragile, always renegotiated – free in one sense and constrained in another. It is a geopolitics without irenicism or angelism, without moral persons, without treaties. In the living world, what ensures limitations on a generalized blind struggle, on the war of all against all, is thus the simple principle according to which communities in an ecology evolve together: every living population is *finite* from the standpoint of its power to subjugate others, and each has a vital need for the powers of many others. There is, indeed, a geopolitics of life: it is hidden in plain sight, behind the big concepts of the ecology of communities, concepts that tend to naturalize phenomena that might just as well be politicized.

This new understanding of coevolution – as a geopolitical force that limits the power of any single species to destroy the others, that forces an aggressor to negotiate and thus reduces its proliferation even while maintaining its presence – has major effects on our understanding of agriculture. The paradox that emerges is simple, and it accounts very well for contemporary deviations on the matter of eliminating pests in agriculture: eradicating bioaggressors by the massive use of inputs simplifies the milieu, kills many more species than the one targeted, and thus destroys the spontaneous arrangements for regulation. As a consequence, it accentuates the proliferation of bioaggressors.

Conversely, the more one accepts the presence *in small proportions* of bioaggressors, and the more one actively encourages biodiversity in an agricultural context, the more one activates the race to disarm, the race to maintain the biotic geopolitics through which species coevolve so as to limit the omnipotence of one to the detriment of the others – that is, to limit its proliferation.

Processionary caterpillars are destroying the pine forests in the Landes region. Why? Because these are simplified ecosystems. For example, tits are among the rare predators of these caterpillars. Now, tits nest in cavities. But because pine plantations exclude standing dead trees and a diversity of other entities that might have cavities to shelter tits, these birds are absent. A plantation is not a forest; it lacks the balancing arrangements allowed by abundant coevolution.

## Realigning Alliances

Proliferation is a pyric power of the living world, and at the same time it is a dynamic that the living world constantly limits through the diversity of species that consume one another, act as parasites to one another, and compete with one another. Agriculture can take inspiration from this double strength. The logic of agriculture or sylviculture thus changes with this approach: the farmers and foresters who practice it throughout the world do not seek to *eradicate* bioaggressors but rather to limit their *proliferation*. And this is what an ecosystem is spontaneously able to do, whereas it does *not* know how to eradicate parasites or predators. Our great confusion comes from the fact that we have missed that nuance: the problem of bioaggressors has never been their *presence* but rather their proliferation, insofar as that proliferation can destroy a harvest and structurally or suddenly weaken the economic viability of an agricultural exploitation. The recent madness of the race to increase agricultural harvests has led to the massive use of pesticides that no longer simply fight proliferation but fight off the very presence of bioaggressors. Milieus are thus simplified, as pesticides kill a broad spectrum of species – meadow birds, for example, some of which limit the proliferation of the insects targeted by pesticides. But a simplified milieu constitutes a catalyst for proliferation of bioaggressors. The passage from the fight against proliferation to the fight against presence produces the opposite effect from the one expected.

The challenge once again is to trust in the dynamics of life: they "know," blindly, how to operate that nuance. This is an established fact. At the Grand Laval farm, knowledge of the fact has led to proactive hospitality for the richest, least utilitarian diversity, simply because diversity works toward collective, multispecies habitability, and thus to our own. No angelism, nevertheless: this attitude, which implies massively minimizing the use of pesticides, is difficult for farmers to adopt. There is something anxiety-producing in renouncing a powerful tool, the death-dealing spreading that gives a feeling of ad hoc, reactive, effective control against an unexpected proliferation. Sébastien Blache often notes how hard it is to learn to be more tolerant toward damaging events, even when they do not make the farm more fragile economically. Doing so requires being well versed in the ecology of communities: it is because we know that the quiet omnipresence of bioaggressors is the norm for an ecosystem, and that overall biodiversity helps to keep such "pests" below the level of proliferation, that this attitude is possible, even if it is hard to maintain. Making the wild powers of the milieu compete in the regulation of bioaggressors makes it possible to accept their multiform presence more easily.[45]

But this also requires a change in mentality among consumers: it is a matter of accepting spotty apples, somewhat bruised fruit; and once again it is an economic problem that implies changes in *supply chains*. In fact, the specifications of long supply chains necessarily spread the requirement that apples be perfect, subjected to exogenous norms concerning their size and appearance, norms that are essentially cosmetic and have nothing to do with the nutritional quality of the fruit. These norms become unsustainable in an agriculture that has changed its relation to wildness. In a cascade, we face the need to valorize short supply chains, to establish circuits with more pluralist norms, to accept less perfect fruit. The short chain with a producers' store practiced by the Grand Laval farm also allows for a form of client education: the farmer can explain to the clients in person why the fruit he is offering is a bit spotted, what higher logic lies behind it all; in so doing, he will be contributing to the change in culture that we need. It is demanding to be a farmer today, when farmers must also teach new and different relations to the living world at the cash register.

## The ambivalence of the "ravage"

Maintaining the presence of a multiplicity of potential bioaggressors does more than simply make it possible to limit their proliferation. The presence of these "despoilers," these parasites who inflict damage on harvests, most often with the purpose of feeding themselves, has additional benefits, not only for the ecosystem and for farms, but also for all of us humans who feed on fruits and vegetables.

On this point, a recent experiment undertaken in the Texas A&M AgriLife Research laboratory, which consisted in piercing small holes in leaves of strawberry plants, enabled scientists to make a decisive advance concerning the very metaphysics of our relations with the plant world.[46] The object of their investigation was to determine whether the products of organic agriculture were better for human health than those of conventional agriculture – a debate that has been driving research for some time now. The answer is in the affirmative, but why? The first intuition is simple: owing to the reduction in the spread of chemical products, the fruits and vegetables would carry fewer of those products, and consequently less of that pollution and less of the damage it causes would be found in our organisms. This intuition is correct, but it is not the essential factor. It is not primarily because we consume fewer pesticides that untreated plants are better for us; it is for a deeper, much stranger reason – one that takes us back to the most distant past, to the moment of the forgotten ascendancy common to plants and ourselves, that mythical time still present and active in our bodies.

What the researchers observed first was that vegetables and fruits resulting from organic agriculture are much richer in *antioxidants* than those resulting from agriculture using inputs.[47] These molecules fight inside our organisms against the oxidative stress that is responsible for cell aging, and in the process they contribute massively to global health. Now, why are these fruits richer in the molecules that defend *us* against illness? Because they have had to defend *themselves*. Defend themselves against whom? Precisely against those agents that intensive agriculture seeks to eradicate, having set them up as absolute agents of harm and labeled them pests, despoilers, destructive organisms. By punching little holes in the leaves of strawberry plants, the researchers showed that the "higher levels of healthy phytochemicals reported in organic fruits and vegetables could be due to the wounding component of the biotic stress attributed to insects to which the plants are exposed."[48] According to Luis Cisneros-Zevallos, a scholar in horticulture and one of the authors of "Solving the Controversy," the responses to the stress created in fruits and vegetables triggered a preharvest increase in antioxidant components.

How are we to understand the contribution of stress? The intriguing phenomenon by means of which the strawberry plant synthesizes antioxidants is tied to a peculiarity in plant forms of life: unlike animals, the lineage of plants did not engage in the adventure of mobility. It experimented instead with another possible path, another way of being alive: unable to move away in the face of threats and aggression, plants became experts in applying alchemy *to themselves*. They are capable of transmuting their own internal matter, transforming the elasticity of their cells to resist wind, raising the level of tannin in their leaves so as to intoxicate herbivores, and *expressing their genomes differently* so as to transform the chemical composition of their fruits. "Several genes related to biosynthesis [of phenolic compounds, i.e., antioxidants] and sugar transport were overexpressed" in the strawberries whose leaves had been stressed, according to Facundo Ibañez: "As a result phenolic compounds and total soluble sugars increased significantly."[49]

## Realigning Alliances

Let us imagine a strawberry in our hands. It is the strawberry plant's need to defend itself against the stress of parasites, against an insect attack, that makes it vigorous and thus able, according to an almost animist motif, to give the strength it has acquired by the ordeal to those who consume it (just as the heart of a bison gives the indigenous hunter his legendary pugnacity, forged by wolves and winter).

The strawberry's wealth in nutrients *for us* comes from this prodigy of the plant world: the fortifying alchemy of its own flesh, as a response to encounters with the external world. Its own defense creates defenses that defend us. Kinship has its obligations, referring us back to our shared ancestry.

It is somewhat troubling, though, to note in the article cited above that the biologists carrying out this study continue to use terms such as "wounding" and "attack" to characterize the actions of organisms that increase the healthfulness of the fruit for us, at the very moment they have discovered the phenomenon. It may be time to imagine terms that do more justice to the *ambivalence* of these organisms toward agricultural practices: they are capable of destroying harvests on the one hand, invigorating them on the other. This is what ecologists call amphibiosis: an interaction between interdependent species that is beneficial in some respects, harmful in others.[50] It is essential, as we have seen, to distinguish between presence and proliferation; here is where the difference between the two effects of parasitic depredators – harmful or beneficial – plays out. Amphibiosis is in effect the hidden name of the broad life of an ecosystem, for, if we vary the scales of time, space, and species, all relations are amphibiotic. For example, in the framework of what are called "durable interactions" in ecology, predation and parasitism are in certain respects *beneficial for the prey and the hosts* alike, as the parasitologist Claude Combes has so convincingly shown.[51] The ambivalence in which we have to navigate is the common basis for all vital interactions; this is why our relations with other living beings call fundamentally for adjusted consideration, and never for war, or for a fantasized peace.[52] It is truly an interspecies diplomacy – always renegotiated, always circumstantial – that this amphibiotic world requires of us.

## *A self-fulfilling prophecy*

In agriculture, "pests" are thus a cause of the nutritional quality of harvests – a complete reversal of the paradigm. But, of course, these are pests maintained below the level of proliferation that would put the farm in danger. Consequently, extractivist agriculture built on the massive use of chemical pesticides, a form of agriculture that seeks to *eradicate* bioaggressors, constitutes an aberration even from the standpoint of its explicit goal of feeding humans: it mortgages the very capacity of harvests to nourish us in the qualitative sense. Without these bioaggressors, fruits and vegetables are poor in nutrients, in flavor, in transmissible strength.

Trust and distrust are not abstract sentiments: these are the names given here to basic practical attitudes that have profound impacts on the world. Distrust has the concrete effect of unraveling the dynamics of life, as we see here in the way extractive agriculture manages the proliferation of "pests." Armed with its ideology of improvement, that agriculture has tirelessly impoverished agrosystems by simplifying rich and diverse ecosystems. In the process, it has dismantled its world and unleashed the aggressions of milieus against themselves: it is because of this cleansing-by-emptying-out that "nature" attacks farming inordinately, in the form of harvest-destroying proliferations. This effect of its own attitude turning against itself *confirms* its philosophy of distrust, which then triggers a vicious circle of suspicion, leading to ever more phytosanitary treatments, and thus to greater simplification of the biotic community (for no pesticide is specific – they always kill more species than they target). Distrust is a self-fulfilling prophecy.

The economic infrastructure that conveys and perfuses this model of relations with the living world is, of course, decisive: in this piece of earth called Europe, the machine that makes such practices durable is called the Common Agricultural Policy (CAP). With its 60 billion euros (about 72 billion dollars) distributed each year, this is the most massive arrangement for international subsidization intended to steer relations to the living world in the agricultural context. The French collective Pour une autre PAC (For Another Common Agricultural Policy) has proposed a luminous critical analysis of the "perverse subsidies" that entail economic favoritism toward the major intensive, mechanized, industrial exploitations. The dominant machinery of the CAP explicitly privileges agricultural forms that fall in the category of what we are calling extractivist agricultural practices: those that are unsustainable. The "For Another Common Agricultural Policy" collective proposes specific reforms and spells out the lines of the coming struggle for a profound transformation of that system. The nub of the problem can be summed up in a single statement: to "change the world" today, we have to change the CAP. Armed with new conceptual compasses for discerning the boundaries between extractivist agriculture and the other forms, we shall rekindle nothing without following this specific local political path.

At the opposite pole from that agricultural model and its "world," there are practices that rely on trust in the dynamics of life: we can look at polyculture, for example, or at the subtle enlistment of wild predators aimed at minimizing "despoilers," or at minimal and well-integrated uses of inputs. This other relation to the living world also has concrete effects on ecosystems, which respond differently. The milieu activates the race toward disarmament on its own, in the geopolitics of multispecies cohabitation that produces a modus vivendi and reduces proliferations, even as it maintains a broad diversity of bioaggressors, but at lower levels, more tolerable for farms. Trust, once again, does not mean laissez-faire, giving in, blissful admiration. It means in-depth knowledge of interdependencies; it means leaning on these, inflecting them with adjusted consideration, and modulating the dynamics of life. Trust is exigent in requiring ecologic intelligence, tolerance, finesse in practices, in design, and in the use of skilled labor. It is not the easy way out: it is the most demanding path, but it is the only sustainable one.

Consequently, one of the great challenges of our time, if farms that are heirs to extractivist agriculture are going to be able to start soon on a massive transition toward these forms of agroecology, is to invent the concrete economic and political arrangements that will make this possible (how can they begin sharing farmlands and struggling against the new great landed estates?), and also to invent new educational arrangements (what sort of agricultural secondary school is needed to change the world?).

## *The diplomatic agroecologies*

There is, finally, one other clue that will help us to learn to distinguish on the ground and in agricultural associations between the practices that belong to extractivist agriculture (which deliberately hides its name), reliant on the invisible metaphysics of production, and the agroecologies sketched in here: it is the dominant epistemological regime – the type of knowledge favored – used to conceptualize the agrosystem and to act on it. This phenomenon is explicit when

*Realigning Alliances*

we examine the agronomic literature that governs practices as diverse as those of the Grand Laval farm, the Agribiodrôme network, the "Project Z" developed by the French National Institute for Agricultural Research (INRA)[53] in Gotheron in the Rhône Valley, along with the traditional models in agronomy.

Extractivist agriculture mobilizes, massively and preferentially, a very particular type of knowledge that comes from a narrowly conceived agricultural science: its common denominator is that it conceptualizes the milieu in terms of categories centered on quantities of matter and energy, articulated according to cost–benefit analyses and vectorized by the productivist issue of yield.[54] Agronomy of this sort is conceived first of all in terms of the quantifiable surface areas, the quantity of inputs used in relation to their costs, the productivity of the biomass, and the final yield: the agrosystem is conceived as a system in which quantities of matter and energy circulate and are spontaneously convertible into monetary value. This way of thinking has advantages in certain contexts; what I am criticizing here is its *autonomization* and its monopolistic tendency. Monopolization is one clue for pinpointing the extractivist agriculture that claims to "produce."

This is not to say that agroecologies, at the opposite pole, care nothing about yields, that they have a purely affective relation to their cultivation; that would be a romantic, immature, and ultimately unjustified opposition. Working with the knowledge base available in agroecology, practitioners continue, of course, to mobilize quantifiable yields, but this is no longer the *dominant* form of knowledge: a different regime of discourse, thought, and science comes to the fore in the approach of these agronomists. The phenomenon is quite spectacular, for example, when we listen to the INRA scientists on the ground who are experimenting with their "Project Zero." What is

## Rekindling Life

at stake is a proof of concept, in which they embark on the adventure of imagining a polycultural orchard whose principle is simple: no pest control products may be used. This project, situated in the framework of an initiative called Écophyto, is just getting under way, but the regimes of knowledge that it has to mobilize in order to imagine this experimental orchard are truly fascinating, if we compare them to the epistemological heritage of the INRA, which has traditionally been more productivist.

Project Z consists in testing a systematic set of alternative practices in the orchard. Let us recall here that tree-growing is a matter of cultivation over a long time span; thus, one cannot bet on crop rotation to limit bioaggressors. This has come to imply, mechanically, using massive quantities of pesticides in order to remain under a rather low threshold of infestation. In this context, the choice of fruit trees to be planted makes all the difference: some varieties are much more sensitive than others to parasites, whereas others actually limit their spread.

To set up an orchard centered on apple trees that will use no pest control products, it is essential to think it through carefully from the outset. The first challenge is to diversify the crops, to get around the fact that homogeneity facilitates the proliferation of parasites. Project Z's experimental orchard, a circular module of 1.6 hectares (about 4 acres) is thus made up of "productive zones" and "production support zones." The imaginary construct of "production" is still omnipresent here in the discourse, but it is already starting to be eclipsed discreetly in the regime of dominant knowledge called upon to conceptualize the system, as we shall see. The challenge is to imagine a multispecies world, the orchard, that limits the risk of pest proliferation at three levels: despoilers must not find it too easy to get in (this is the role, as we shall, of an external tree barrier); they must not be able to install themselves in vast numbers (this is the purpose of selecting diverse varieties); and they must not be able to develop disproportionately (this is the point, for example, of associating apple trees with other fruit trees, which allows what is called in ecoepidemiology a "dilution effect").

Around the orchard, then, a circular hedge of trees constitutes a barrier, a windbreak, and "room and board," in the words of Sylvaine Simon, one of the architects of Project Z. Some of these trees also produce harvestable fruit (chestnuts, almonds, hazelnuts); their flowering seasons are spread throughout the year, with local features that have coevolved with the indigenous insects.

Let us take a particular pest as an example, the rosy apple aphid, and experience the arrangement from the insect's point of view, trying to grasp the types of knowledge mobilized. The rosy apple aphid, a major parasite on apples, has its own life history: a singular cycle in the common space and time of the milieu. It arrives in the fall in passive flight. Now, the circular hedge has been designed to create a kind of *whirlwind*, and the aphid will thus be reduced to landing on the *first* row of apple trees in the orchard; this is why this row contains a particular variety of apples, one that is not ideal for "production" but that has fascinating properties for the project we are observing. Rosy apple aphids lay their eggs in the cracks of the bark, but when they hatch in the spring they *cannot* develop on this variety; the Florina cultivar inhibits hatching for reasons that remain unclear. This first line of fruit trees is thus an impasse that will prevent the aphids from drifting toward the other concentric rows of trees in the interior of the orchard; it will thus keep them from proliferating. These inner rows are made up of various fruit trees and different varieties of apple trees; they are also interspersed with grasses and bushes that do not "produce" but that shelter alternative prey and flowers for the

## Rekindling Life

auxiliaries in their dizzying multiplicity. At the center of the circle, a small wild ecosystem combines piles of rocks to welcome snakes, a pond to attract amphibians, and piles of wood to shelter mammals; it is fenced off so the tranquility of animal life will never be disturbed by passing humans: this space is left, in a sense, in *free evolution*.

With Project Z, we are far from the bucolic image according to which it suffices to plant trees and wait for the fruit to fall, as in the productivist image of rows of homogeneous apple trees boosted by chemical products, envisioned solely from the standpoint of the tons of biomass of fruit that they generate.

Here is the discreet but decisive difference between the two approaches: knowledge of classical agronomy is predominantly mechanistic, physicochemical, quantitative, in the Liebig tradition.[55] It deliberately valorizes abstract laws governing interchangeable milieus.[56] Knowledge of the agroecologies sketched out here does not entirely avoid these approaches, but it has developed, in addition, a heightened attention to *another* style of knowledge. It takes in, first of all, knowledge of a *multispecies ethology*, investigating *how everything behaves*. One has to know how trees negotiate with the wind to make whirlwinds, to know the agenda of the aphids from the inside, to learn how every tree reacts to aggression, so as to shape this kind of arrangement. One has to know the lifestyle of every manner of living being from the inside if one wants to favor a modus vivendi without resorting to phytosanitary eradication.

Michel Jay, a researcher in agronomy, is a fount of science concerning this agricultural knowledge,[57] whose originality lies in an approach that weaves ethology, as a science of animal behavior (traditionally centered on relations between fellow creatures of the *same* species), together with the ecology of communities as the science of relations *between* species – two sciences that are normally separate in programs of scientific research. Listening to Jay, one cannot help being struck by the difference in worlds between what he sees and describes in an agrosystem, in relation to the flows of fertilizer, tons of seeds, and the mathematical models that populate the imagination of the most quantitative agronomists. Jay explains, for example, that bats have, as part of their evolutionary inheritance, the ability to regulate their resting temperature, in keeping with that of the milieu. This implies in practical terms that the ideal shelter for a bat varies in relation to temperature variations: it is thus necessary to place several shelters on a farm to harbor these mammals in sufficient numbers and let them activate their dynamics.

But this hybridization between two sciences is also enriched by a particular philosophical attitude toward these life forms: that of a virtual traveler capable of decoding the customs of foreign peoples – foreign and yet kin – who cohabit with us. This agroecology is thus *diplomatic*, because it has transformed in its very practices the ontological conception of the nature of life forms other than our own: we can now see them as peoples with strange, exotic customs. We need to know how they use the world, what they require, what their habits are, in detail. This attitude is quite visible in Michel Jay's approach, for example. It is a matter of asking oneself questions like these: How do blue tits feed on insects? Which ones do they prefer? What do they like and what do they find repugnant? What frightens them, or attracts their curiosity? How many trips do they make each day between their nests and their prey? This questioning approach requires learning to see from the viewpoint of the multiple life forms present so as to negotiate the appropriate modus vivendi. One might thus call this regime of agronomic knowledge an *agroethology*.

Thus, the nature of the dominant knowledge is what differentiates extractivist agriculture from the agroecologies we are considering:

schematically, the first approach can be characterized as a mechanistic agronomy that functions according to a simplified view of physicochemical systems; the second, as a diplomatic agroethology that reckons with the geopolitical complexity of the amphibiotic relations among life forms.

This difference points to a broader nuance between objectivizing knowledge and diplomatic knowledge: in the latter approach, non-human living beings are no longer consigned to the old nature of the Moderns. Tracking the proportion of these two types of knowledge, these two attitudes, brought to bear on a given farm makes it possible to understand with more precision to which family it belongs in the intricate continuum of practices (since these are of course two ideal types, and each farm combines in different proportions practices situated at different points along the continuum).

But the difference goes even further: as we have seen, the quantitative approach of extractivist agronomy translates biological quantities into economic quantities *without remainder*. It is the search for a common cost–benefit metric that makes it possible to put such disparate factors as bushels of apples, the ecosystemic services rendered by wild biodiversity, work hours, and the cost of each pound of fertilizer into a single equation. Diplomatic agroecologies have a different relation to the milieu/profit equation: they recognize that, if it is indeed necessary to quantify yield and determine the economic viability of a farm, the biological processes at work cannot be fully translated into economic terms. There is an intrinsic *incommensurability* between what a farm brings in, what the farmer does, and what the milieu does.[58] The farm needs to be economically viable – this goes without saying. But many practices – setting up nesting boxes, activating ecological dynamics, experimenting with forms of hospitality for wildlife, enjoying concrete effects from the reinvigorating of the site – are not factored into the economic equation; it is simply a matter of inhabiting a place and taking care of the milieu that takes care of us. This is another decisive clue that helps to identify such places.

What is confounding, when we circulate in the non-mystic mandala of Project Z, is the density of intelligence per square yard: intelligence of the scholars woven into the different but active intelligences of the living beings present.

And, of course, a diplomat is in a sense a gardener – but gardeners come in two sorts. The first arrange things according to their own viewpoints, without taking the time for close observation of the forces at work; thus, they simplify the milieu in a self-reflecting way. Wendell Berry sums this up in a luminous nuance: the American colonizers certainly approached the New World to cultivate the land with a vision, but they lacked *sight*.[59]

The other sort of gardener, the one who presides over the alternative practices we are investigating, is found among scientists and farmers alike. Their shared approach amounts to observing ecosystems from the viewpoint of interdependencies, with a precise and original goal: allowing new ideas and practices to arise from the inventivity of life itself. Multiplying what they see in order to shape their own vision.

The point of these agroecological projects is to put agriculture at the service of biodiversity, and to put living diversity at the service of agriculture. The originality of such projects lies in this chiasmus; it goes beyond the dualism that contrasts the utilitarian approach (putting wild auxiliaries at the service of exploitation) and the approach of the traditional preserve (renouncing all exploitation to the benefit of wild life).

The challenge lies in determining how to favor natural regulations, how to lean on "nature" to prevent the explosion of bioaggressor populations and to preserve the fertility of the soil while still maintaining "production." Here, again, we recognize an attitude characteristic of trust in the dynamics of life: minimizing substitution, refusing to interfere with the functionalities at work when there is no need to do so – refusing, for example, to smear clay on tree trunks, thereby letting wild helpers do their work.

These, then, are actions that encourage the functionalities of the milieu instead of substituting for them; actions that reconstitute dynamics instead of cutting them short; actions that allow interrelations to express themselves instead of deactivating them.

Can natural regulation generate significant yields? This is the problematics of Project Z: one can see that it retains certain aspects of INRA's productivist legacy. This is what makes it ambiguous, while also giving it the strength to convince the agricultural world. This is the fascinating hybrid question, a blend of modern (thus ancient) productivism and diplomatic agroecology, raised by the experimental project begun in February 2018; at the time of this writing, it had not yet offered up its first fruits. It will be several years before we have answers.

## *Returning to the common enemy*

Let us now take a broader perspective, moving to a higher level of philosophical generality and swooping down from there onto the political stakes of these questions. Our philosophical reinterpretation of extractivist agriculture as based on a distrust that devalues the dynamics of life allows us henceforth to understand, by contrast, the philosophical identity of the agroecologies concerned with "feeding humanity and healing the earth" (of which permaculture is a recent form).[60] From the foregoing inquiry, we can deduce a common denominator for the various agroecologies we have examined: they are fundamentally based on active trust in the dynamics of life.

Consequently, we understand why the opposition between exploiting and sanctuarizing is a metaphysical legacy that creates toxic political effects in the way it distributes friends and enemies, where the uses of the earth are concerned. If this dualism corresponded to reality, diplomatic agroecologies and extractivist agriculture ought to belong to the *same* camp: the one determined to produce food, *against* all forms of sanctuarization. But if the essence of the agroecologies lies in trust in the dynamics of life and struggle against the ideology of "improvement," we can see how absurd it is to include them in the camp of extractivist agriculture.

And if, finally – especially if – that same trust *also* underlies the combat and the conviction of those who support hearths of free evolution, those who defend wild nature, it becomes clearly apparent that a *mistake* has been made about where the border between allies and enemies is situated.

If we choose the wrong criteria to understand them, agroecology and free evolution are opposed to one another. If we choose more carefully, they become links in the same chain of efforts to reinvigorate milieus. Exploitation and unsustainable exploitation, free evolution and sanctuarization, must not be confused. "Must not be confused": this is a key phrase in philosophy – it can prevent us from drawing up inaccurate maps featuring false borders that send us straight toward the abyss, headed for disaster. Multiplying the maps, adjusting them to the stakes, makes it possible for us to open up paths for action.

The line between "trust" and "devaluation" thus constitutes a new political frontier. It passes henceforth between extractivist agriculture, allied to all the unsustainable uses of the earth, on one side, and, on the other, all the agroecologies, allied to strong protection of milieus, to free evolution, and even to rewilding, in a non-dualist sense, in a coalition among *all* the sustainable and life-enhancing uses of milieus.

If the opposing camps are beginning to be drawn differently owing to the trust/distrust operator, we must explore further to see how trust can be translated in *practical* terms. This is not a vain wish or a fuzzy sentiment. It is not a declaration of intent offered to agrobusiness, one that could be printed on billboards as a way of "greenwashing" any sort of practice. Trust and devaluation: each of these attitudes logically implies *a style of action* opposed to the other, a style that can be identified in the field on the basis of precise criteria. How can one recognize trust at ground level? It is manifested by a quite precise style of action – one that calls for minimizing substitutions, stimulating the existing dynamics, supporting the available functionalities, letting the milieu regenerate itself.

But how can we assert that trust implies a style of unified action if we are attributing that style to free evolution – which consists in prohibiting any intervention, any act of adjustment and exploitation – and at the same time to the agroecologies that are, on the contrary, very intensively involved in action of that sort? There seems to be a paradox here. If we are starting to see an emerging alliance between free evolution and agroecology, in defiance of all the traditional dualist classifications, we can interpret this as a philosophical symptom of the fact that the opposition between exploiting and sanctuarizing is not the *only* unfounded dualist opposition we have inherited. What this improbable alliance demands of us now is that we deconstruct the dualist opposition between acting and "letting nature take its course," between transforming domestic nature and contemplating wild nature. The notion of trust is a compass that allows us to reconstitute our dualist conceptual maps, one by one.

*Realigning Alliances*

## The question of action in conservation biology

Since trust implies a style of action, we cannot conceive of conservation biology as a radical rejection of all action. To grasp what is at stake here, we need to make a short detour by way of the debates that are animating communities of protectors of nature. In these milieus, the dualist heritage is quite visible in the form of the latent opposition between contemplation, which is valorized, and action, which is often viewed as intrinsically destructive. It is this opposition that puts conservationists in an awkward position when they are asked to name what they do. The dualist specter underlies the debate over tagging birds, for example (they are tagged to protect the species – but isn't tagging them already doing violence to their wild nature, sometimes even weakening their rate of reproduction?). The question arises in more general debates over monitoring wildlife (camera traps help us to better understand fauna – but are they not symptoms of a desire to know everything, to control everything?).

The paradox that brings the untenable character of this opposition between acting and sanctuarizing to the surface is that those seeking to preserve nature are constantly obliged to *act* in order to sanctuarize particular milieus: by arranging habitats in order to save an endangered species, by removing pollutants, by reintroducing elements that had disappeared from the site. These are indeed all actions. How do we escape from this chaos? The problem is that this formulation concentrates all the attention on the attitudes of humans (must we act or not?), even when the issue is defending something else: the good health of a milieu, or of a species. Concentrating solely on the question of human attitudes (should we transform or contemplate?) localizes the problem incorrectly: our focus should be not on ourselves but on the dynamics of life. On adjusted consideration for a milieu, for example, so it can regain its own functionalities. Sometimes this implies a hands-off stance, and sometimes it implies intervening so the milieu's inherent functionalities can take over again. The problem is not defending what is *intact* (we have gone beyond the backward-looking dualist dynamics of *wilderness*); it is not that human action is condemned as unnatural and thus opposed to wildness (humans are living beings among others): action and non-action both serve the ambition of allowing the dynamics of the milieu and its evolutionary potentials to be expressed.

The challenge here is to leave behind the misanthropic demonizing of action on milieus and of its symmetrical sanctuarizing. It all comes down to a concrete, delicate, circumstantial analysis of the ecology of a milieu, of its biological heritage and of human interventions. It is in these troubled waters, far from any abstract principle of management, that conservation professionals have to navigate, and this is what makes their job so hard.

What concrete actions can be taken in conservation biology, then, to restore to the dynamics of life the possibility of expressing themselves? For one thing, useless dams can be eliminated, in order to allow the functionalities of a wild river to be reconstituted. For another, a designated milieu can be protected from any harvesting or exploitation. Nesting boxes can be installed in an organic tree farm to *allow* tits to *come back*. In the light of examples like these, we see the originality of such actions: their goal is to eliminate their own necessity, so that they need not be maintained, tended, or repeated. They work toward their own disappearance. For the vocation of such actions is to reactivate the dynamics that surpass us, and allow them to get back to work. This is why conservation, so active on the ground, seems always to valorize contemplation of nature over exploitation: contemplation can be enjoyed once we have undertaken the minimal action required so that the milieu no longer needs us. This is what happens with what is called "lending a hand" in conservation biology: lending a hand means carrying out a one-time action that does not need to be maintained by active management over the long run – an action that allows the spontaneous return of species, relations, and functionalities (for example, the reintroduction of indigenous species capable of redeploying on their own). It can also take the form of subtractive management, when man-made obstacles, "anthropic forcings," are removed (the elimination of unnecessary dams is the most powerful example here). The distinguishing feature of this type of action with respect to the spontaneous dynamics of life is not that we are acting *in their place*, but that we are leaving them room to act *on their own*. Free evolution, which seems to do nothing, is thus paradoxically an action of this sort.[61]

## Toward coalition

Now that we have dispensed with the opposition between action and non-action in the field of nature conservation, we can deduce what this implies at another level, in the relation between the protection of milieus and agriculture. The idea is simple: there is a profound link between the type of action that takes place on its own in free evolution, in the removal of dams (the old "protection of nature"), and in agroecologies. These practices are not identical, of course, since they differ in their goals: in one case, protecting a milieu, and in the other, producing a consumable biomass to feed humans. But, while they must not be conflated, they must certainly not be viewed as opposites, either. They should be seen as two complementary faces of a shared relation to the living world. What free evolution and an agroecology hospitable to wildlife have in common is a shared *style* of action that is based, as we have seen, on trust. This action launches and orchestrates the dynamics of life. It allows these dynamics to express themselves; it does not oppose them, or substitute for them (or it does so as little as possible). Free evolution manifests this same style of action while allowing a milieu to be relieved of all forms of exploitation; the *spirit* is the same.

What is interesting about this compass is also that it makes it possible, in the field of productive activities, to clarify the distinction between sustainable and unsustainable exploitations: this border distinguishes a style of human action that activates, prolongs, and respects the dynamics of life from a style that opposes, mutilates, and substitutes for those dynamics.

Profiling the problem in terms of styles of action with respect to the dynamics of life thus makes it possible to configure the plural field of uses of the earth differently. It brings indiscriminately into the same camp heterogeneous practices and relationships to the earth that find themselves opposed to one another and irreconcilable in terms of our inherited archaic oppositions between wild and domestic, intact and arranged, exploitation and sanctuarization, improvement and laissez-faire.

Through all this, a new coalition is emerging in France. It was initially unified by a common enemy. But that was not enough, for the unity was reactive. Hearths of free evolution, conservation without "managitis" (the drive to amend everything in order to protect given milieus), alternative forestries, and sustainable agricultures (like the organic polyculture-stockbreeding in the Grand Laval farm) – all these practices are very different in their ends and means, but they are located on a *single continuum* of political alliances and communities of thought. In addition to a common enemy, they have a stronger bond that makes this continuum a *family*. Some members may be very far away from others in this family, to the point of believing themselves to be on opposing sides, or in conflict on the local level, but each remains allied to the others, once what all members share has become clear.

The common denominator of this coalition has three key components: a shared spirit of trust in the dynamics of the wild; a shared project focused on rekindling the embers of life and restoring autonomy to particular milieus; and a shared outlook embodied in the quest for adjusted consideration.

Nevertheless, we must not erase the differences between the goals of "conserving nature" and those of the agroecologies; these are two quite distinct expressions of the same trust. Hearths of free evolution affirm trust without the aim of cultivation; they are neutral from the standpoint of subsistence. They make sense more or less anywhere on a territory, in relatively small but significant proportions to the agricultural lands devoted to subsistence. They are compatible with the types of agroecologies defined above; these latter constitute a different expression of trust – one that is inflected, put to work, designed first of all to feed us.

Here, then, is our new coalition. It has already come about in numerous places, without much fanfare, on the ground in France, but it remains in the minority and invisible. Transforming exploitation in the direction of better care for the living milieu in its plurality, agroecological in inspiration, is beginning to make sense for many farmers, when they have sufficient economic and philosophical margins for maneuver to leave behind the legacy of "agricultural modernization."

An interesting example of this coalition has developed between a land user and the ASPAS project Réserves de Vie Sauvage. A winegrower in Gervans, in the Drôme region, has decided to transform his operation by creating a number of habitats for wildlife in his vineyard. And, at the same time, he has decided to give the money earned from the sales of a vintage rosé wine, the result of an especially favorable season, to the Réserves de Vie Sauvage project. When he was asked to explain his decision, he replied that he wanted "to thank Nature for the recent fine season, and to apologize for the harm he had done [to Nature] in producing that wine." It is not a matter of "compensation" in the economic sense of the term. Put in everyday terms, economic logic amounts to something like "you pay for what you break, and if you can pay, you can go ahead and break things." Here, the winemaker's gift is not offered as a replacement for transforming his agricultural practice toward more sustainability through a shift to organic viticulture and a commitment to wild diversity in his vineyards. He characterized his donation rather as a counter-gift, a form of reciprocity for the gifts he has received from the living milieu. He imagines his act as a means, as he put it, "to apologize for the wrongs that intensive agriculture has inflicted on Nature." He has adopted a relation to the milieu that no longer pits domestic against wild, exploiting against taking care – a relation far removed from that of agricultural modernization. This vintner has freed himself from the metaphysics of production, which rejects the reciprocity to which our giving environments commit us. And yet he is neither antimodern nor premodern; he is sketching out the premises of a desirable future. Agroecologies – understood from here on in the sense of agricultural practices that no longer purport to be "producing" subsistence, that no longer devalue the gift of the dynamics of life, and that are committed to various forms of reciprocity (amendment of soils, reinvigoration of milieus, defense of wild biodiversity) – can thus constitute a figure of the future for an agriculture liberated from the metaphysics of production. These agroecologies are already thriving on a different map of life.

*Realigning Alliances*

The members of this coalition share a sensitivity to the plurality of life forms, to the different ways of being alive. A sensitivity to relations. A capacity to see from the standpoint of interdependencies, and to bring them into the field of collective attention. In sum, they share a sense of adjusted consideration. This is embodied in a spectacular way by naturalist farmers. These fascinating characters, obsessed with the living world, are able to name every species; they can decode the ecological dynamics carried out by these species on their lands, and they see the ecosystem in terms of interdependencies.[62] The same sensitivity is also embodied in the defenders of free rivers and forests, people with intimate knowledge of the non-mutilated capacities of the milieus in question. Letting a forest age, letting a river flow in both directions (with sediments shifting downstream and fish swimming upstream to hatch) – practices such as these exemplify adjusted consideration for a milieu.

Since the health of milieus takes many different forms, there is no single way to behave with adjusted consideration toward a given milieu. This fact underlies various projects for *transforming land use* toward more free evolution (and stronger protections in general), and more sustainable uses (agroecologies, polyculture combined with stockbreeding, short supply chains, agricultures that care for the soil). It is no longer a question of opposing the two approaches, but of rethinking their *proportion* and their *distribution*.

We can now understand why the defense of milieus in free evolution must not be viewed as a way of defending something that is intact, as a radical refusal of all human action, for to see it that way would mean disconnecting it from the *other ways of rekindling the embers of life*. Free evolution must be conceived in different terms: as the extreme limit of a continuous spectrum, and not as a unique, radical, exceptional type of protection of milieus that turns all "less pure" approaches into enemies and thus ultimately condemns itself to failure. There are countless forms of trust in the dynamics of life all along the continuum that goes from free evolution toward agroecologies and diplomatic forestries.

Bringing to light the possibility of such a coalition among scattered actors – in France, the Network for Forest Alternatives (RAF), the Wildlife Reserves (RVS), the Naturalist Farmers ("Paysans de nature"), the agroecologists, the nonviolent sylviculturalists, the Drôme Farmers' Confederation, and dozens of others – has a quite specific political function: to keep the groups from exhausting themselves in fratricidal conflicts like the ones that beleaguered Leninist Trotskyites and Trotskyite Leninists – while, on the other side, the destroyers of the living world rub their hands in glee at the sight of so many internal divisions.

"Know who are your allies and who are your real enemies": this is the watchword for the new ontological map that should make it possible to orient ourselves properly in the conflictual space of politics.

It takes the form of a question to put to ourselves. Why am I defending a preserve? Is it out of latent misanthropy and fidelity to a cult of the intact, or out of trust in the dynamics of life as experienced by a living being, the dynamics that weave a common world?

Why am I defending small-scale farming? Is it out of fidelity to a cult of cultivating a milieu against a deficient and hostile nature, or as a way of rejecting the industrialization that does violence to the dynamics of life and to the humans who depend on it?

This distinction really does constitute a conceptual map, not a prophecy or a vain wish. The political effect it seeks comes down to calling for in-depth alliances, as opposed to superficial and circumstantial alliances and enmities. In-depth alliances are, by definition, submerged, hard to grasp at first glance – and yet they are the most

## Realigning Alliances

genuine of alliances. This is the wish behind the new map: that those who defend relations animated by trust in the dynamics of life recognize one another and form a common front. That the defenders of alternative forestry and the diplomatic agroecologists recognize free evolution as their ally. And that the defenders of integral bioreserves recognize these forms of exploitation not as faces of Exploitation stigmatized wholesale (unless they are prepared to stop using wood, and stop eating), but as allies in the defense of the embers of life.

These actors must weave themselves together in a coalition that is open to internal discord, but not to the extent of weakening the struggle against their real enemies: this is the diplomatic rule I am proposing here if we are to form a common front.[63]

Here, we can grasp simultaneously the conceptual motifs discovered in the course of this inquiry: rediscovered trust in the dynamics of life (implying adjusted consideration, and reciprocity toward the milieu that supports those dynamics) constitutes a compass for imagining thousands of wide-ranging levers of ecological action, a coalition of actors producing a confluence of actions designed to rekindle the embers of life.

### Interspecies alliances

If we push our exploration of these in-depth alliances to the limit, there is one that emerges in a decisive way: the interspecies alliance between human and nonhuman living beings.

Pollinators destroyed by the massive use of chemical inputs, fauna in soils depleted by extractivist agriculture, and so many others – these living beings come to our attention today in a new way. Inasmuch as these beings are essential elements in the action loops that make our lives possible, by weakening them we are weakening our own conditions for life. They call themselves to our attention as means toward our ends, means we are destroying, and thus they become visible as *something other* than means. In the process, they point us in new directions, of which they are objective allies. For example, the pollinators in the French countryside are currently declining in massive numbers – whereas their activity is what brings fruits, vegetables, and all flowering plants, wild and domestic, back to us every spring. If we listen carefully to what their distress reveals, it becomes clear that they are being poisoned by unsustainable uses of the earth (pesticides, ecofragmentation, simplification of ecosystems, destruction of hedges, and so on). In other words, these living beings – bees and other pollinators – tell us what we are doing wrong: bees point their antennae toward the agricultural practices that are destructive to the milieu that we *share* with the other living beings. Without awareness or intent, but with a very sure sense of what is vital, they designate the dividing line between friends and enemies regarding the uses of the earth. They show us what to change in order to make the world more inhabitable, for us *and* for all the other living beings, without distinction. A bee that indicates to an entire society that its manner of doing agriculture is destroying the entire milieu, a bee that reminds us of the wholesome ways of dealing with the earth: is that not a creature that deserves to enter into the fabric of our common world, into the field of political attention – and with the status of ally?

We must, nevertheless, distinguish from the start between these interspecies alliances and a form of instrumentalization: for example, we could speak superficially of an alliance between a farm and a wild species when the spontaneous action of that species offers services to the farmer. However, quite often such a scenario reflects only a utilitarian sense of "alliance": it entails putting a form of wildlife to work because that work is useful to humans. The sense defended here goes well beyond this instrumental sense. It is not a matter of claiming that, because these nonhuman beings serve the exploitation of a territory by humans, they deserve to live, or to be respected; the question of "deserving to live" is not the heart of the problem, since we are not arguing in the moral realm here but rather in the political realm, with the antagonisms proper to it. It is a matter, rather, of showing how alliances with species with which cohabitation is complicated are not alliances contracted out of duty or for utilitarian purposes. We must show, instead, that they make sense inasmuch as they valorize *trajectories leading to a transformation* of the way we use the land, a transformation toward practices that are unstinting toward *the relationship* between more emancipatory human actions and ecosystems as a whole. Such alliances will help us to better understand that what destroys the living world ecologically is also toxic for the conditions of human existence. Seen in this light, alienation is a trans-species phenomenon.[64]

In the face of the current crisis in our relations to the living world, a crisis threatening the very conditions of human life, it is not arbitrary to *hypothesize* that the arrangements that alienate humans are often the same as those that alienate nonhumans. The human activities that assign no value to cohabitation in their quest for short-term economic profits also give no priority to emancipating the workers at the heart of an enterprise, or to increasing access to more flourishing forms of life. These activities unfold to the detriment of the living conditions of all the actors involved, humans and nonhumans alike. Any activity that implies as a matter of principle the need to destroy or devalue some part of the milieu that is vital for the worker can hardly claim to be liberating for that worker. No agriculture that destroys the life of the soils – in particular, through the massive use of inputs – can claim to be truly liberating for farmers, who are often its first victims, as one can infer from the social and psychological effects of the "green revolution" on the rural world. It is an ecopsychological paradox that we are confronting here. The human activities that involve nonhumans are all, without exception, facing the alternative of being conceived either in terms of complex and fragile partnerships with living beings, or else in terms of control of a biotic community reified as matter, a form of control based on coupling maximal exploitation with the eradication of pests. Yet, and here is the paradox, I am arguing here that *despite* its ideological justification as rational progress, the choice of the second axis, if it is destructive for a swath of inhabitants of the territory, is inevitably also alienating for the actors in the practice themselves.

Consequently, it can be argued that the forms of practice that are intrinsically diplomatic toward the living world are more spontaneously liberating and gratifying for the actors and human communities that apply them. It is the intriguing happiness of permaculture practitioners and diplomatic agroecologists (like the happiness of those whose food they supply) that is the enigma here. These farmers are at a vast remove from the Adamic paradigm of fighting to civilize the earth by means of a plow, killing the enemies of the harvest, and earning one's bread by the sweat of one's brow; instead, they flow peacefully into delicate alliances with what they cultivate, seeking to transform what others see as pests into helpers. This does not mean that everything is rosy: agents harmful to certain crops exist; parasitism and predation exist. It is simply the way a farmer relates to these phenomena that changes, and thus also the associated agricultural and economic practices. If those harmful agents are understood as curses inflicted on us, whose destiny as humans is to maximize production, they are harmful in themselves; if they are understood as cohabiting partners caught up along with us in delicate and complex

## Realigning Alliances

political relations, with whom we must invent alliances, minimize or deflect discord, polish multiple relations, convert competitions into mutualisms, then we pass from a state of war and exploitation to a state of complex alliances with the agroecological community – alliances more sustainable for all parties. Agroecologists consistently claim that the essential aspect of their trade amounts to shifting from a logic of adversity to a logic of partnership with the living world. I am arguing here that this partnership is an art of effective liberation on the part of the practitioners themselves. In the modes of distribution, a short-circuit supply chain is bound up with this objective alliance, which weaves together a large number of actors: producers and consumers and living beings on a much broader scale.

As a result, these vital alliances with nonhumans, with wild dynamics, are woven together with uses of the territory that, in the current context, are often *simultaneously* more viable for the biotic communities involved and for human activities in what is *human* about them (that is, for the workers' conditions of existence, as well as for the *meaning* workers can give to their work).

With such alliances in place, we are no longer defending living beings against humans in general, against all human uses, in order to sanctuarize them, to set them up as irreducible ends in themselves (even if it is sometimes necessary to do that); political ecology defends living beings as allies of the best practices applied to the earth, allies of "trajectories for transforming" the entire technological system toward uses of the earth that are more sustainable, more respectful, less wasteful of energy, and more resilient. Best practices mean more delicate uses, endowed with adjusted consideration. This is the new watchword for acting politically in the twenty-first century, for there are only interdependencies, and the only decent politics of interdependencies entails unending negotiation of adjusted consideration, in opposition to all unsustainable practices. Here, politics does not mean a parliament of the species, but a collective engagement weaving together humans and nonhumans in trajectories for transforming the uses of the world.

This concept of interspecies alliances, like the others proposed here, is a map: its aim is not to provide an exhaustive description of reality but, rather, to allow us to orient ourselves in a way that is more fluid, more vigorous, better adjusted. Still, countless other maps remain to be invented by cartographers, for each one is circumstantial, and its only value lies in its power to shine a light on life, in order to open up paths for action.

In these vital alliances, what is living in us is allied with what is living outside of us; what matters to us and to the others proves to be indistinguishable.

Nevertheless, this hypothesis has to be empirically tested, time after time. I am defending it here as a tendency, not as a necessity or as an a priori truth: with respect to the extractivist appearances of current Western productive activities, it is highly probable that, in many cases, uses of the earth more concerned with cohabitation will serve the transition of territories toward practices that are ecologically more sustainable and *humanly* more livable, in unison. Of course, in hundreds of particular instances, given the stubbornly complex character of reality, this will not be the case: the whole problem comes down to identifying the places where cohabitation with wild lives is opening a path toward what is more sustainable for human *and* nonhuman communities, and toward forms of alliance and negotiation. Perhaps this orientation, once clarified, can help to liberate the imagination of the actors involved as they seek desirable trajectories of transformation for our uses of the earth.

5

# Making Maps Differently: Dealing with Disagreements

*Creating a sanctuary means establishing good connections*

The tools forged by this investigation allow us to take a new look at some of the current controversies over uses of the earth, controversies that continue to pit protectors of nature against farmers and foresters. I shall now try to reformulate the problems differently in the light of the new maps sketched out here.

To do this, it will be useful to begin by reconsidering the new status acquired by what our tradition has called "wildlife preserves." If the binary opposition between exploiting and sanctuarizing no longer holds up, it means that the way we think about strong protection – that is, about the creation or maintenance of sanctuaries – must also be reconceived in a more nuanced way than simply as the polar opposite of any human activity. We have established that it is not human activity in general that is destructive of the living world, but, rather, human uses of the earth that are based on antagonism toward the dynamics of life (manifested in distrust and substitution). In this framework, sanctuarizing can no longer be defined solely as setting aside a given land area, keeping it free from any exploitation, following the model Edward O. Wilson has offered in *Half-Earth:* set aside 50 percent for biodiversity, 50 percent for exploitation of the earth.[1] At first glance, Wilson's proposal may appear powerful, its simplicity and apparent radicality appeasing our feelings of impotence and guilt with regard to our legacy of destruction: here at last is a measure that makes room for life. The weakness of the proposal, however, is that it has not adequately examined its inheritance: it is wholly constructed around a blind spot.

As environmental history and ecological thinking show clearly, the dualist logic behind nature reserves shares a history with Indian reservations in North America.[2] This logic, applied to wild fauna or to indigenous Amerindian peoples, justifies situations in which a dominant group appropriates land for new uses deemed "productive," with a corresponding expropriation of "others" whose presence is limited to the small spaces assigned to them. This is a logic of fencing in or fencing off, confining the "others" who formerly lived *everywhere*, to exceptional spaces designated as "sanctuaries" – which are, in fact, ghettos.

The dualist preserve reactivates this logic in neocolonial forms. We cannot defend life in Western Europe while neglecting the legacy of our colonial history, an important component of which is the inequality of our ecologic and economic exchanges with formerly colonized countries.[3] We cannot defend the idea, for example, that the major part of French or European territory must become a territory in free evolution; this would be to accept the idea that our food has to come from elsewhere, presumably from our former colonies, via low-cost economic supply chains that would simply prolong the colonial relationships and would end up destroying their ecosystems to the benefit of ours. Our care for the dynamics of the living world has to be maintained both at home and abroad, simultaneously. And, to achieve this, we have to reestablish autonomous food supplies on our own local territories. A small percentage of well-connected hearths of free evolution, spread more or less everywhere throughout the landscape, surrounded by agroecologies and traditional farming freed from dependency on the capitalism of agrobusiness, along with nonviolent sylviculture: this is the picture of what we need.

But the most serious flaw in the logic of *Half-Earth* is that it maintains the dichotomy between humans and nature, and consequently the dichotomy between sanctuarizing and exploiting; in so doing, it validates the claim that exploiting necessarily means exploiting unsustainably.[4] This is the paradox of the traditional dualist approach to nature reserves: it presupposes the destruction, outside the reserves, of what it seeks to protect within. This is evident in the American tradition of conservation, intrinsically bound up with the history of US capitalism: a wilderness preserve constitutes by its nature something like the small percentage of clear conscience that allows capitalist agrobusiness to exploit all the rest blindly. In other words, the problem of the old logic of sanctuarizing to which *Half-Earth* belongs is not what it does inside the sanctuaries, it is what it permits and justifies elsewhere. The problem is not what it cherishes, but what it neglects.

Wilson would certainly deny that the sanctuarization he has proposed legitimates unsustainable exploitation, but his claim is based on a postulate that no one could find credible – one that constitutes the blind spot and the weak point of his entire proposal. To imagine that human production and human activity could be relocated onto 50 percent of the land on earth, Wilson has to appeal to a *deus ex machina*: he postulates a technoscientific solution, thanks to which humanity will miraculously invent, in the coming decades, ways of providing food and producing energy that could satisfy a *growing* human population, while exploiting only half of the earth's land. He does not lay out a concrete path for the "brilliant" inventions that would allow us to believe in this technological dream. More humans fed, housed, and laundered by half as much exploited land: this is what reveals that *Half-Earth* is built on sand. This magic solutionism allows the revival of the dualist dream. But that dream will not become a reality.

The constitutive problem of the dualist idea of a preserve, a sanctuary, is thus that it functions as whitewashing – or greenwashing – for destruction everywhere else. The same society that has destroyed milieus has now invented blind exploitation and sanctuarization as the salve to its conscience. But the problems, as Albert Einstein used to say, cannot be resolved by the models of thought that led to them. In other words, the challenge for an association interested in strong protection of milieus is to offer a new model of relationship with the living world – a more mature and fruitful model than that of the traditional dualist preserve – while maintaining the goal of strong protection. This is the line along the summit that is so difficult to find.

The challenge is to invent a new concept of preserve – one that transcends the exploitation/sanctuarization dichotomy. This is what I am trying to do here through the concept of hearths of free evolution articulated with the rest of the territory: the dualist approach has to be surpassed by a concept that reverses its logic. Hearths of free evolution are not what comes *after* extractivist exploitation, in reaction, as compensation, as a way of saving the furniture while accepting the destruction of everything else; they offer the original compass for sustainable relations with the earth. They are no longer "for nature" as opposed to "for humans"; they are hearths radiant with life for the community of living beings, a community to which humans belong. They are hearths that connect with other uses in the surrounding area, uses that are also nonviolent and sustainable – uses allied with the hearths against unsustainable exploitation. These other uses would also be categorized as exploitation, but they would be endowed with adjusted consideration toward the powers and requirements of each milieu, inasmuch as any given milieu has the capacity to regenerate on its own.

For hearths of free evolution to mature, they must always be conceptualized in a chain of alternatives in relations with the milieu. The situation has changed since the time of John Muir and the conservationist logic of American-style preserves: today, we can no longer defend nature in free evolution, or wild animals, without at the same time defending the kinds of alternatives in the human uses of territories that make these cohabitations possible. Many associations for the protection of nature are, in fact, beginning to understand this.[5] In concrete terms, it means that an association for the protection of nature can no longer defend wildness without conceiving of a worldwide social project. We can no longer protect nature apart from the soil, as if there were no world, as if we were not consumers. We can no longer "protect wild nature" without defending a compatible human world that would be fulfilling for relationships among all forms of life.

A preserve conceived as a hearth of free evolution whispers to the world that we must no longer accept the idea that the earth is fated to see 99 percent of its land exploited blindly and 1 percent sanctuarized; we must fight for a reversal of that destiny – a scenario in which the earth is exploited in a way that is sustainable and compatible with wild dynamics on the vast majority of its territories, left in free evolution elsewhere, and protected in myriad ways more or less everywhere. The old camps constituted by the old approach to strong protection – misanthropic ecologists against destructive exploiters – no longer hold up: the enemy of strong protection is no longer exploitation as such, but unsustainable exploitation. Hearths of free evolution are the allies of all uses of the earth that are respectful of the living world;

they breathe with that world, they conspire with it to imagine other uses of the earth – good uses, finally, that benefit all parties.[6]

## *Returning to the forest beyond dualisms*

We can now solve the problem with which the first part of this book concluded: How can we simultaneously conceptualize different relations to forests – some involving exploitation, others free evolution – without setting them in opposition to one another on the basis of dualist categories? What type of integrated forest management does our new logic allow us to imagine?

If we come back to the case of the ASPAS reserves, the association does not demand, for example, that free evolution monopolize the management of all forests. The idea is not to transform all forest ranges into forests in free evolution, where harvesting and exploitation would be banned. Hearths of free evolution are designed to be articulated with other forest spaces where other types of management deserve to be imposed, on the basis of trust in forest dynamics as a style of action – these spaces need to be articulated with sustainable farming, with irregular hedges, multispecies populations, management respectful of the logics of forests, as in the Pro Silva model: alternative forms of nuanced diplomatic management, as in the Network for Forest Alternatives;[7] permaculture forest-gardens; inhabited forests in which harvesting uses are maintained (gathering wild edibles, collecting mushrooms and wood). This continuum must be envisioned as a gamut of uses and relations. Human relationships with forests can take many forms, including practices of exploitation that deserve to be defended, provided they are sustainable. There are non-exploitative uses that imply nothing more than harvesting (gathering herbs, for example). But free evolution is *also* a specific use, a weaving-together with a forest: we go into a forest to immerse ourselves in the lives of other beings, without leaving any traces. Free evolution implies an intense relationship between a forest and the human world, because it constitutes a fountain of life that flows around and toward the other forest forms, invigorating them.

Hearths of free evolution are thus one alternative *among others*, situated in a very specific camp: the family of sustainable relations with the forest, as opposed to its cheapening. This family groups together an alliance of alternative forest management approaches whose common denominator entails defending adjusted consideration toward the intimate logics of forest milieus. Most importantly, these alternatives are unified politically by a common cause: resisting and struggling against *unsustainable* exploitations of forests and milieus. Thus, a dialogue has begun in France between ASPAS and RAF concerning possible bridges between the two approaches, and about ways of protecting forests whose owners would like to maintain sustainable "soft" forms of exploitation.

What compass can we use to imagine a sustainable exploitation of a forest? This is what RAF, the Pro Silva management, and all nonviolent sylvicultures have understood: such exploitation will be based on the same powers of life, regeneration, and resilience found in ecosystems in free evolution. These ecosystems do not serve as norms dictating what every forest milieu *must* be. But they indicate the type of eco-evolutionary processes that have the greatest potential for sustainability, diversity, and resilience. Sustainable sylviculture takes its inspiration from these ecosystems: it is regenerative, it practices proactive hospitality for its dynamics, and it limits its impact on wild biodiversity.

We must thus understand hearths of free evolution by repositioning them in an integrated conception of our *plural* relations to forests.

For what is good for a given milieu is not unique and univocal. Forests show us different ways of relating to them, different possible uses that are all endowed with adjusted consideration for what a forest is (uses that are sustainable and respectful of forest dynamics), because ecosystems are rich in potential trajectories. For a forest, for an ecosystem whose dynamics are still autonomous, what is good can be expressed in multiple ways. This is what might be called the *polymorphous health* of an ecosystem or a life-supporting milieu. In contrast, what is bad for a milieu is often clear and univocal (reduction of resilience, of diversity, of adaptive potential, of functionalities).

There is an analogy with humans here: what it means to mistreat humans is fairly clear and univocal (it is more or less the same for everyone), but, conversely, the ways in which each person can flourish are multiple, even though they are adjusted to each person's conditions and form of life.

It is this polymorphous, plural character of the health of milieus that opens up a middle way between two classic positions. The first is the one according to which an ecosystem does not call for adjusted consideration because nothing but matter is involved. Nothing is good or bad in and of itself, independent of humans. Nothing is better or worse for the ecosystem, only for us humans, who are endowed with a monopoly over the privilege of attributing value. As an aggregate of matter, an ecosystem lacks any immanent normativity. This "modern" approach has served to justify blind exploitation, devoid of all respect toward the milieu, which is viewed as a simple storehouse of resources. According to the second position, we are limited, when we want to manifest respect toward a forest, to a single type of dogmatic management, on the grounds that a thriving forest has a single unchanging type of vitality and good health (for example, free evolution, absolute sanctuarization). Both of these approaches are backward-looking legacies from dualism; they miss the heart of the problem. An ecosystem whose dynamics remain autonomous nevertheless has several potential trajectories in store that are rich with adjusted consideration, that maintain its adaptive potential, and that are distinct from the extractivist trajectories that would weaken, degrade, and eventually destroy it. This is why it is essential to multiply alternatives that are sustainable and attentive to adjusted consideration toward forests, in an approach that does justice to our possible relations with a milieu. The concept of the polymorphous health of a milieu is an important tool for holding together the need for a plurality of uses of forests – implying forms of harvesting, exploitation, housing, and free evolution – while keeping this pluralism from swerving into relativism (meaning that all uses would be acceptable, since no single one would be the absolute norm).

If we, nevertheless, look for a tool of minimal orientation for navigating toward what is good from the standpoint of forests themselves, and adopt a position of adjusted consideration toward them, we have to adopt a forest's perspective – in other words, we have to learn to see from the viewpoint of the interdependencies that make up a forest. This is a difficult art, mastered by ecologists and foresters who have proven capable of this type of decentering and can see milieus on a long-term scale. One minimal marker is precisely whether or not we can have confidence over the long term, since, in the living world, the long term is always a temporal regime that creates a more adjusted modus vivendi for the species and dynamics that are present, from the angle of coevolutions. (The short term, in an ecosystem, is often a more conflictual regime, involving invasion, competition, parasitisms by new pathogens, while the long term, by definition, smooths over these interactions, since natural selection retains only the elements best adjusted to a balance of strengths.) Coevolution is often a reliable guide to what is good for a particular milieu: the presence of a given species, a given dynamic, a given disruption, is good for this milieu because mutual adjustments have been made throughout a history so ancient that the components do not function fully unless they work together, are woven together – this is a minimal way of evaluating what may be good for a milieu.

Imagining an integrated vision of relations with forests is crucial, because it entails realigning alliances: when we hear again and again that free evolution amounts to putting a forest under a bell-jar with the goal of protecting the forest itself against *all* human uses, we are in fact confronted by an ideological argument, massively adopted by the industrial exploiters of wood in order to turn the defenders of sustainable alternatives in forestry (approaches that valorize multi-use cohabitation) *against* strong protection of nature, presented as misanthropic and as excluding humans from the milieu. People submit readily to this rhetoric, and the defenders of more ecological forest management find themselves shoved into the camp of extractivist industry by a trick of language. The same thing happens when people hear that strong protection of nature would defend forests for themselves, against humans; this is an error in reasoning, for hearths of free evolution in fact defend rich *relations* between forests and humans – in the form of uses of the forest that do not involve harvesting or exploitation, of course, but that are of great value to both parties nonetheless.

Here we need to shift the relentless operation of this inquiry to the precise context of forests – to do this means realigning political alliances in a more nuanced way. The enemies of sustainable forests are exploiters who have inherited "improvement" as an absolute principle, those who reduce forest ranges to lumber mills while exhausting the humus, weakening the soils, and reducing millions of acres of forest land to industrial plantations devoid of birds, leaving the land incapable of regenerating and reinvigorating the surrounding territory. These forests are monocultures that must be fed with pesticides and fertilizer to produce wood by the cubic yard.[8]

The real alliance that deserves to be woven lies elsewhere: it comes about among those who defend diplomatic relations with forests, those who are concerned with interdependencies. And this alliance is woven *against* unsustainable relations.

It remains clear, in the current context, that the majority of France's forests can continue to be exploited as soon as exploitation becomes sustainable. Those in free evolution, in much smaller numbers, have to be one of the faces of that sustainability. All the alternatives can struggle together against "bad forestry" and the destruction of the fabric of life, of life's diversity.

We need to note, however, that this integrated map of forested lands has a differentiated temporal logic: every site has a different relation to time. The forests in free evolution have to be allowed centuries to develop, while the most exploited forests have temporalities that are shorter but still require attention to the cycles proper to the particular forests themselves, in relation to their ecological structure.

If an entire forest range were like a body, each hearth of free evolution would be like a heart, as powerful as the heart of a blue whale, streaming life all around, toward territories where management is active but sustainable. The forests defended by RAF and Pro Silva would be the muscles, which produce biomass on which one can draw in a respectful manner for the regeneration of milieus.

If an entire forest range were like a watershed, wild forests would be like multiple springs that flow and nourish the water table. Permaculture forest-gardens would be little canals inflected to produce fruit and plants for humans, taking their inspiration from the strength of forests while allowing them to regenerate.

Rekindling the embers in all milieus, and in all directions. The debate is open – and it will be conflictual – over the place to allot to spaces in free evolution, the place to grant to milieus offered for harvesting (wood and mushrooms, for example, where forests are concerned), the place devoted to nonviolent sylviculture, to inhabited forests; but there are no solutions apart from an *integration* of these different land uses based on trust in the dynamics of life, in a coalition that will activate an inch-by-inch struggle against other uses of the forest – those that devalue, destroy, or structurally weaken the fabric of life. This is an appeal for a systematic transformation of human uses of forests toward greater sustainability and toward adjusted consideration for the nonhuman life forms that populate these milieus.

The destiny of the earth, according to the metaphysics of improvement, was to be exploited rationally everywhere, so that exploitation would ultimately be valorized as the only good use of the earth. Here, in contrast, I am defending a multiplicity of good uses of the earth: an alliance between intensely ecological exploitation, conservation of milieus, free evolution, rewilding, on a continuum of trust in the dynamics of life and in adjusted consideration, *against all unsustainable uses*.

The most urgent need is indeed to protect our habitats – that is, the habitats of living beings. Our habitat is that of all living beings: the hearth of one form of life is only the weaving together of all the others.

To protect a form of life is to protect one's world. This is true for human and nonhuman living beings alike. And, in a stroke of luck – countering all those who seek to set the causes to be defended in opposition to one another– it turns out that the world in question is one and the same.

## Rewilding beyond the dualist myths

If we want to take the concept of alliance to its extreme limit, we must confront another debate that promises to be stormy: how to maintain, at one and the same time, the defense of cultivation and farmlands *and* the defense of a return to wild dynamics in free evolution, when the two vectors *seem* to be diametrically opposed.

The form this debate takes today in rural areas is often open and intense conflict around the idea of "rewilding." This notion crystallizes an opposition concerning the future of the land. Rewilding is one of those words that "sing more than they speak," as Paul Valéry would say – a word that has more emotional charge than precise definitions, a word that triggers visceral reactions of enthusiasm or rejection.

Some hear in this term a redemptive cult of the Wild – others, a dangerous misanthropy; some see it as a return to an innocent past – others, as a regression toward a mythic Eden that has never existed.

Rewilding is indeed the precise technical term for certain programs in conservation biology.[9] Nevertheless, there is confusion owing to the fact that the concept has several meanings, which its affective connotations too easily conflate.

Rewilding consists in reinvigorating and protecting the dynamics of life and of the biotope, through three types of actions, sometimes deployed jointly, sometimes separately: preserving hubs of nature in free evolution; ensuring connectivity among these hubs; and reintroducing key species.[10] In the face of milieus in which human activity has mutilated certain ecological dynamics to the point that they have been *simplified* and have lost resilience, rewilding consists, for example, in reintroducing species capable of activating ecological functions so as to restore complexity to the milieu, and it consists in reconstituting the equilibrium (for example, between predators and prey) that had been developed by now-destroyed coevolutions. At the outset, these projects may take recourse to forms of ecologic engineering (for example, to reestablish connectivity between hub zones or to reintroduce species where they have disappeared), but the goal sought is to favor self-regulated and self-sufficient ecosystems that will require no management, or very little. The reduction or complete absence of human management is thus the sign of success for a rewilding project. Free evolution is one of the modes of managing rewilding. And, in a sense, it is the ideal for long-term management of any successful rewilding project (often an unattainable ideal, but that does not matter – every swatch of autonomy restored to a particular milieu already produces important effects).

Here we have a minimal technical definition of rewilding. The majority of the points in question deserve to be debated in relation to each context. But what generates confusion, fundamentally, among the different meanings of rewilding are the explicit goals adopted by the various initiatives. For the same tools for managing a milieu can serve several different ends. And it is ultimately the philosophical conception of the relation between humans and nonhumans that governs these differing aims, which are often not made explicit. To take one example, a significant portion of what is called Pleistocene rewilding seeks to reintroduce populations of wild animals corresponding, more or less closely, to Paleolithic megafauna. But the project of reintroducing bison into mixed forests of central Europe can have two different goals. The first could be characterized as reactive: a return to Paleolithic landscapes inasmuch as they embody a form of intactness (in the face of a misanthropic observation that the world has been intrinsically polluted by human presence), or a return to a mythical prelapsarian time (that is, to a time before humans began to destroy everything). However, that same project need not be reactive, misanthropic, or backward-looking: reintroducing bison may aim at reinvigorating the dynamics of the forest milieu, by reconstituting coevolutions that have played an important role as stabilizing arrangements for the ecological trajectories of the milieu. The meaning of a rewilding initiative depends on its philosophical positioning, and this positioning has major practical implications for the management of the ecosystem, but also – and especially – for the relations that these initiatives maintain with the human collectivities directly concerned with the project. The social acceptability of this sort of project is at the heart of the matter – without it, any such project, no matter how ecologically relevant, will be doomed to fail.

We can thus distinguish three types of rewilding, starting from the often unacknowledged philosophical stance of each type regarding the place of humans in "nature." There is, first of all, the rewilding that I call misanthropic. It postulates that all human action and human presence intrinsically constitute pollution; it rests, paradoxically, on an absolute dualism according to which humans are of a *different* nature from "nature." It has simply reversed the stigma: in modern dualism, this human difference is a choice that elevates us above the nonhuman; here, it is a curse on the order of Adam's fall. From this standpoint, the goal of rewilding is to reconstitute intact milieus and to reject all human presence and activity.

The rewilding I call backward-looking rests for its part on the idea that the normative state of particular milieus, intrinsically good, is the one that ecosystems manifested *before* the Neolithic revolution, with that era's implications of cultivation of milieus and demographic growth. It is based on the strange anthropology according to which the only form of humanity that is of the *same* nature as "nature" is the pre-Neolithic form (depicted, or fantasized, as hunter-gatherers living "in harmony" with nature). The intriguing idea behind this conception of humanity is thus that the Neolithic era, as the turning point at the dawn of civilization, generated an ontological difference in the humans who are its heirs: the very nature of those humans was changed. The fascinating paradox of this anthropology is that it adopts once again on its own account the worst aspects of the Moderns' colonial anthropology, which consisted in distinguishing between civilized humans, who were *not* part of nature, and "savages," those wild humans perceived as ontologically closer to wild animals than to civilized humans – an arrangement plainly serving to justify slavery, expropriation, spoliation, and colonization. But here again the conception of the human that underlies this backward-looking rewilding has reversed the stigma. It depicts civilized persons as an intrinsically impure and culpable form of humanity, portraying the ancient "savages" as the only innocent and defensible forms of humanity, because they were not separate from "nature." From this standpoint, the norm for ecosystems is established as the way they were before the founding myth of civilization appeared – the myth of Neolithic agropastoralism.

Both of these positions are simply reactive reversals of modern dualism. They share much more with this dualism than they acknowledge: claiming to triumph over it, they are in fact simply prolonging it.

Of course, no rewilding project rests explicitly or monolithically on either of the two anthropologies we have just examined: the often disproportionate part those anthropologies play in concrete projects of rewilding takes, rather, the form of impulses, visceral reactions, unexamined feelings. We have to ferret out these specters of indefensible anthropologies in the language and practices of rewilding in order to determine their relevance, critique them when they are misguided, and reorient their energies.

How should we think about rewilding in the light of the idea of trust in the dynamics of life? What I have learned from my inquiry allows me to depict a third type of rewilding, built on another idea of the place of humans in the living world.

In the approach I am defending here, there is no cult of intact nature, and no desire to return to the Paleolithic – each particular milieu has inherited a history that weaves together geology, climate, evolution, and human action. It is not a matter of wanting to bring milieus back to the state they were in prior to human action, but rather of allowing them *from now on* to develop according to their autonomous dynamics without being sliced up: free rivers, without unnecessary dams, that are able to breathe; forests that are allowed to age and radiate with lives, without being "harvested" in their adolescence. And this would take place with their past and its exploitations as a starting point: it does not matter whether or not these sites are "intact," or what their trajectory has been – the interesting question is where they are *headed*, once they have been freed and allowed to express their inherent strengths.

It is simply a matter of giving them the space and the time to express themselves.

This rewilding is built apart from the dualist map that functions in the zero-sum mode: rewilding does not act for "nature" to the detriment of humans, but rather for the community of living beings of which humans are members. In this sense, we might call it "solidarity rewilding."

## Rekindling Life

This solidarity rewilding, then, consists in supporting the regeneration of the properties of the milieu that *function on their own*.

According to Gilbert Cochet's definition, in rewilding "there is nothing to *do* in the strict sense." It is enough to *undo* what has been done and what has mutilated the vital ecological dynamics (unnecessary dams, for instance); and to *redo* what yesterday's damage has destroyed – thanks, for example, to the "helping hand" actions that have reintroduced eradicated species, such as beavers, who work on rivers, or vultures, who are the great healers of milieus, purifying them by eating carcasses and by neutralizing pathogens that their own bodies have the power to digest.

In rewilding, we are not regenerating life – that is not in our power, frankly, as we have seen; instead, we are priming life's autonomous powers of regeneration. We allow life to express its own resilience. We put in place the minimal, delicate, distinct conditions for it to regain its full vitality.

In this precise sense, rewilding is not a purist position, "alone against all the rest"; it is a link in the chain of protection of milieus, allied with sustainable forms of forest and agricultural exploitation, and more thoughtfully developed forms of conservation.

### *Defusing the conflict between rewilding and the rural world*

In the rural world, the opposition to rewilding is often formulated as a defense of farmlands and farming culture against the takeover of territories by rewilding, and the practice of rewilding is often caricatured as a form of financialized capitalism aiming to turn virgin lands into sanctuaries. As we have seen, however, these charges do not apply to the forms of rewilding supported here. Moreover, this conflict mobilizes sociological caricatures depicted as actors: "the greens" pitted against "the farmers." It tends to crystallize around the issue of wolves, for example, or the issue of wildlife preserves, or the issue of pest management. The conflict is often invisible to city-dwellers, but it has been quietly structuring the debate in France over the future of the land in a large number of rural areas, especially in the mountains.

Why does the rural world seem dead set against rewilding? Some, in that world, sense beneath the term a cult of Wildness, a defense of wildness against "civilization by way of the plow," a defense of "letting go" as opposed to managing. In other words, they sense a challenge to a certain centuries-old interpretation of the rural mission – that of cultivating the earth to extirpate it from its original savagery. Rewilding is thus associated with an idea of regression. But we now understand that the notion of cultivating the land as a civilizing act is an ideological product of the metaphysics of improvement – in France, it can trace its origins back to the figure of a pioneering bishop who managed to fulfill the earth's destiny by transforming an insalubrious swamp into agricultural land. While it is true beyond doubt that cultivation has made it possible to feed populations and to clean up milieus that had been vectors of disease, it is once again the elevation of that improvement into an absolute, overriding principle – its claim to monopoly – that has to be criticized.

The conflicts between the rural world and "the greens" have other motifs as well, most notably sociological in character, folded into the gaps between the lifestyle of the neo-rurals who have come to stay, often under the impetus of globalized "ecological" discourse, and that of the local residents who have never left. But what crystallizes this opposition and makes it intractable, in my view, is the dualist legacy in language and attitudes.

## Rekindling Life

We have already seen the various wellsprings of this dualism. What is interesting here is the way it works in either/or, zero-sum terms: in dualism, everything that is for the wild is against the domestic, and vice versa; everything that is given to the one is taken away from the other.

This dualist legacy is the basis for a strong antagonism between the rural world and "the greens," an antagonism that paves the way for extractivist exploitation. The latter seeks to divide, the better to conquer – sometimes leaning on an alliance among farmers against "the greens" (for example, the common front of FNSEA and the Confédération paysanne on the matter of wolves), sometimes mobilizing, with distortions, the slogans of the defenders of natural milieus.

To get beyond the impasse, we shall need to learn to think in terms of interdependencies between rewilding and rural agroecology. As we have seen, as soon as we have operated a two-fold, better-adjusted redescription of what is in play, the dualist conflict falls away on its own: on the one side, we have dismissed the misanthropic dualist conception of rewilding, which the rural world legitimately criticizes, in favor of "solidarity rewilding," which does not oppose all human activity – instead, it serves all parties, by supporting the return of the autonomous dynamics of the living world that have been mutilated by unsustainable exploitation. On the other side, we have seen that sustainable farming is based on the same dynamics of life and on the same trust in them that characterizes rewilding. Consequently, solidarity rewilding and sustainable rural farming – agriculture based on trust – are allies against all forms of industrial exploitation of the dynamics of life.

Still, moving beyond dualisms in the matter of land use does not allow us to resolve all conflicts magically in good-natured harmony. It allows us to do something else, though: we can rid these conflicts of

*Making Maps Differently*

their philosophical trappings, strip them down to their real, residual motives. And we can do this with two goals in mind: on the one hand, we can try to shift the front lines by engaging with the real adversaries and avoid creating unnecessary enemies; on the other hand, we can try to ferret out the real motives behind conflicts hidden under lofty words. Let me be clear on this point: once the specter of dualism has been exorcized, the remaining conflict between the rural world and the rewilding project will be over power relations regarding real estate – the struggle over ownership and land use. To whom does the land belong, and in what proportions? Is it a matter of sharing space? It seems to me that sharing is a defensible approach, although not in the form of dualist zoning: rather, in integrated blocs, with little hubs of free evolution (like hedges or ponds on a traditional farm), in decent proportions (the RAF proposed 10 percent in its forest ranges), connected by ecological corridors, but also well woven together with the other uses of the milieu – linked to all the other sustainable, traditional, intensely ecological and sociologically emancipatory exploitations in the vicinity.

The question of land distribution is thus the remaining point of friction, the turbulent zone in the dialogue and in concrete local compromises. Above and beyond struggles between private actors, the driving force behind this bone of contention can also turn it into a fertile controversy: a collective, democratic debate over the *trajectories of transformation* that each territory wants for itself. But it is no longer appropriate to cloak the conflict in the noble philosophical garments of confrontation between two "worlds," two irreconcilable value systems – this would be dishonest. The struggle is now being waged on other grounds.

This does not mean that, in the light of the approach I am proposing, a harmonious agreement will reign henceforth. The controversies between the rural world and the world of conservationists – how to distribute the spaces to be exploited sustainably and those to be set aside for preservation – will remain current, but they will play out within a single camp: in a coalition that will be capable, moreover, of making common cause against all unsustainable exploitation.

Such an alliance may already be under construction in France by certain local actors. For instance, in the course of their dialogue, ASPAS and the Drôme Farmers' Confederation have succeeded in acknowledging that their real enemy, their common enemy, is in fact the artificialization of the soil, a practice that compromises the maintenance of rural lands and crops as well as the defense of ecosystems.[11] Similarly, the decisive effects of the destruction of rural milieus and of old-style farming practices are combined in large-scale industrial farming and extractivist monocultures that rely on petroleum and on unrestrained mechanization. Moving beyond dualism in the uses of the earth takes place through simple and concrete actions, recognizable in the fact that they are backed up by other dividing lines. ASPAS, for example, is currently reflecting on how best to manage a situation such as the following: what to do when donors bequeath land suited for farming to the association so that it will protect wildlife on the site? In most such cases, rewilding would make no sense. The preferred path would consist in co-management of such spaces, in which farming associations that are struggling for access to land would collaborate with farmers committed to a form of agriculture that concretely activates trust in the dynamics of life (in France, the association Terre de liens – part of the European network Access to Land – is one possible model).[12] This type of collaboration, even on the local level, is crucial for sealing the alliance between rewilding and intensively ecological agriculture.

## *Making Maps Differently*

To combat both extractivist exploitation and dualist sanctuarization, the renewal of small-scale farming and the pursuit of rewilding are complementary projects that need to be supported, in order to rekindle the embers of life.[13]

# 6

# Conclusion: The Living World Defends Itself

This book is centered on the living world as a world to be defended. But there will be objections. Why stop with the living world, why not go further? Why not defend and preserve the universe, the earthly crust and kernel, clouds, the climate, every molecule in the waters of the ocean? Why prioritize living beings above all else, and not packets of atoms? Why does the community of those to be defended begin with the family of living beings?

Because living beings are the first beings for which something matters.[1] In the cosmos, the emergence of living beings is the great ontological revolution: before it, there were only stones, forces, and gasses that were not interested in their own existence. A planet devoid of life would not warrant protection, because what lacks life cannot suffer. Nothing matters to a pebble, a photon, or an atom – nothing matters to any thing that exists without life. It is because beings for which something matters appear that there are reasons to begin our action of care and defense with those beings. The appearance of life in the known cosmos some 3.8 billion years ago inaugurated the existence of beings for which certain things have importance, have value – living beings, from bacteria to whales. With living beings, then, *importance* is born. We living human beings are certainly packets of molecules, but with a sense of importance shared with all the other living beings.[2] A pebble is indifferent to where it is lodged, whether it is in your shoe, on a mountain-top, or in outer space, while

## Conclusion: The Living World Defends Itself

the tiniest bacterium is concerned with its location, which it actively configures as a milieu. It has a better and a worse position (in individual and populational terms). Gasses and molecular compounds do not experience stress; it has been well documented, however, that plants do.[3] Theirs is different from the stress human mammals experience, but it is a related form – at one and the same time alien and kin. And it hardly matters whether the bacterium or the plant is "conscious" of this condition for its own existence, or whether it "knows" anything (these are inoperative anthropomorphic concepts here) – these beings are sensitive to their milieus, they can *prosper* there, bringing about the tiny difference that makes all the difference. This is our originality as living beings in the cosmos, 99.999 percent of whose mass is constituted by packets of inorganic molecules for which nothing matters. And it is because something matters that there is a world to be defended, and reasons to struggle. If we share our atomic material with the stars (from which our molecules derive), we share a *family* with the other living beings – that is, a common ascendancy, a mutual vulnerability, and a shared destiny.

It is astounding to observe that, even today, with its wings trimmed in the countable category of biodiversity, the living world is passed off in politics as a secondary question, childish, unserious. The living world must be defended because it is important: it is *importance itself*, since it is the very source of importance in the universe.

Nevertheless, we must not take the term "living" in the strict sense to exclude mountains, rivers, soils: after all, the inorganic material of mountains is made from ancient skeletons of living beings (and it comes, still further back, from the stars); the same water circulates from rivers into the blood of our arteries; the hard wood of trees is the carbon in the air. Inorganic matter circulates in the living metabolism called the biosphere. Inorganic milieus are merged with living beings: they are constitutive of life. The dynamics of life, defended in this book, are precisely the forces that make abiotic elements circulate in the living world: they are like the oxygen that animates a fire, feeds and modulates its potentialities. And it is as habitats, as biotopes, as elements taken up in this metabolic circulation of living beings (in ecology, these are called "abiotic conditions"), that rocks, rivers, and the climate are implicated imperceptibly in this struggle and this preservation. Inorganic matter constitutes communities of living beings through these relations: they are not apart from the soil, but woven from air, climates, stone, and water. The slogan "Save the Climate" is thus an unfortunate metonymy: the climate risks nothing. We living beings are what need to be preserved from disruptions of the climate; thus, we need to throw our strengths into the struggle against climate change. The abiotic world is woven together with the living world in an imperceptible way, but in this weaving, living beings emerge with a particular quality: they are *concerned* by what is happening to them. This is their originality in the cosmos, their ontological specificity.[4]

Taking the living world as the unit to be defended radically transforms what is commonly called "protecting nature." Protecting "nature" has been, up to now, the preoccupation of a small number of people – the "greens," the "friends of animals." Beyond those circles, up to now, this was of interest to very few, and for good reason: "nature," according to our dualist legacy, is everything that is not us, that is, the inverse of us humans – everything that exists apart from us, and that we pollute just by existing. Why should we feel concerned, then? Protecting nature belongs to the realm of idiosyncratic passions – there have been people crazy about nature, as there are people crazy about cars, or postage stamps.

But what becomes of the protection of nature when we have understood that what has to be defended is not "nature" defined in

## Conclusion: The Living World Defends Itself

dualist terms: not what is intact (as opposed to exploited); not what is pure and "natural" (as opposed to artificial); not what is wild (as opposed to domestic); not what is remote (as opposed to *domus*); not what is outside, symmetrical to us humans in general? What does protection of nature look like when we take the living world as a unit to be defended? *Who* then becomes concerned, and in what forms?

### The specter that haunts us

Nobody really believes in the human/nature dualism, but its specter haunts us. It does not govern our private lives or our experiences, but it is activated whenever we come into conflict; whenever we speak (for it has monopolized words); whenever we make our oppositions absolute; whenever we set up a hierarchy without paying attention to the ambivalences of reality; whenever we are bothered by some otherness, unnerved by some uncertainty. This dualism is not a metaphysics in the strong sense, it is not a world system: it is an arrangement of power configured to enable humans to win through binary exclusion, in the face of the world's complexities and ambiguities – an arrangement that allows us not to see from the standpoint of interdependencies, that allows us to hunker down in one camp against another, to oppose our interests to another's in such a way as to find ourselves winners or victims. As Claude Lévi-Strauss puts it, the nature/culture dualism is not a "concrete aspect of universal order. Rather, it should be seen as an artificial creation of culture, a protective rampart thrown up around it" because human culture felt too weak, at risk of being swallowed up.[5] This dualism entails a rigidification of thought, an absolutized expression of fear. Following Lévi-Strauss's lead, we need to read it from an ethological rather than a metaphysical standpoint: we find in it the impulse to fix things in place, to attribute too much importance to their differences, for fear of being endangered by the otherness to which we assign a minority status. The dualism targeted here is this little ethological phenomenon, institutionalized as a metaphysical category, in a society. That is what makes it so intractable.

One might argue that contrasting extractivist agriculture to agro-ecologies that trust in the dynamics of life – as I have done in this book – reinstitutes a dualism, since we are dealing with a distinction between two categories. But getting away from dualisms does not mean giving up all distinctions (that would mean giving up thought); it means understanding distinctions differently: as solutions to problems in particular circumstances. Dualism in the philosophical sense is much more than a conceptual difference between two things; it is a residue from an ancient distinction, made absolute. It is a hypostasis of two opposed blocs in conflict. It is an incitation to war, with mutual exclusion, and a hierarchical reification of an ancient, local distinction that, in other circumstances, may well have served a purpose.[6] The artificial/natural distinction does serve a purpose when it is not made an absolute and caught up in an exclusive hierarchical dualism within a zero-sum system, just as the fine term "wild" is useful if it is just a calm descriptor. It is not a question of banishing words, then, or the word "nature" in general; it is just *one use* of the word "nature" that must be banished – just the use that is charged with meanings and implications in the dualism opposing nature to humans, or nature to culture.

One might think that getting beyond dualism amounts to entering into monism (a unified metaphysics according to which everything is of the same nature): "We are all Nature." But this is not at all the philosophical gesture being made here. To exit from dualism, in this case, is not to adopt monism, because we are not engaged in a metaphysical debate over the nature of reality – rather, we are engaged in a political debate (in the noble sense of the word "political") over the massive categories we use to operate and activate power relations with regard to what is important for a collectivity. Beyond dualism, there is not some vague, oceanic monism – there are simply more finely drawn categories, endowed with a sense of context; there are non-essentialized dualisms that are taken on as context-specific solutions to explicit problems, and not compartments made absolute as truths in order to arrange the world in two boxes, one of which is an ontological trash can (nature or culture, body or mind, wildness or artificiality).

It is sometimes necessary to deploy a metaphysical nuance in order to shift the lines of a political conflict. If it in fact matters to move beyond dualism, if it is urgent to change our map, it is because the dualist map provokes vital blockages, aporias in collective action, while the new map seeks to unblock and reconfigure the world in ways that would allow us to intervene more effectively on the political and practical levels. The fundamental challenge here is to show how a metaphysical argument (consisting in the move beyond the human/nature dualism by positing the operator "living" as the common denominator) makes it possible to resolve the problem of unnecessary enemies, and thus to redistribute political alliances in relation to the uses of the earth. Abolishing the dualist map on which we stand, changing the cartography of allies and enemies, thus constitutes a political gesture starting from an act of philosophical work on the ground.

## *From human beings to living beings*

Let us go back to the question, then: what does it mean to protect nature, once we have left the human/nature dualism behind? It

## Rekindling Life

means rekindling the embers of life: struggling to restore their vitality and full expression to the dynamics of eco-evolution. Rekindling the embers of life is simply a redescription of what was once called "protecting nature," when we have moved beyond "nature," protection, and dualism. But it is a redescription that necessarily reconfigures practices. We have seen that this formula binds together three philosophical stakes; it is time to recall them here, at the end of this reflection, now that we are fully prepared to grasp the import and meaning of the last point. First of all, life is not a cathedral, it is a fire. Secondly, we cannot protect in a paternalistic fashion something so much bigger than we are – we can only restore to it the conditions of its own autonomous regeneration. Finally, it is not as humans that we protect an otherness known as nature, it is as living beings that we defend the living world – that is, our multispecies milieus.

Retaining "humans" as a unit for designating those who defend their world, in fact, poses a problem. It means bringing back, in spite of ourselves, the precise dualism that has contributed to the fact that we do not feel ourselves concerned with the living world. We have observed this at work on several occasions, moreover, in certain mainstream contemporary ecological positions that stem from a persistent anthropocentrism: these approaches defend the importance of human protection of nature, but nature remains conceived as a reservoir of resources that need to be preserved and managed more reasonably to avoid mortgaging our survival. The poles have not changed: it is simply a matter of moving toward more prudent management for future generations. In persistent anthropocentrism, the dualism is not transcended, it is simply amended.

The inertia of this dualism can also be seen in the field of "nature conservation." The philosopher Virginie Maris argues, for example, that the concept of nature as "otherness" is the ontological category

## Conclusion: The Living World Defends Itself

best suited to designate wild nature. Her theoretical project has the great merit of truly breaking with the modern dualism, but at the same time she maintains a dualist separation into two blocs: humans / wild nature.[7] As we have seen, however, it is not necessary to remobilize an ontological dualism to justify defending milieus in free evolution, from which exploitation and harvesting would be prohibited. Prolonging the dualism may even seem dangerous in the light of its toxic legacy. And, in my view, it is philosophically disproportionate to reactivate the dualism in order to legitimize the existence of nature reserves and the protection of wild places.

If what needs to be transcended is the dualism itself, we cannot do without a philosophical anthropology – in other words, we cannot avoid raising the question: what are humans in relation to the rest of the living world? For, if the other living beings are indeed other, "alter," they are also our *kin*. It is a matter of maintaining simultaneously their right to be what they are, different and autonomous life forms, and the fact that they are not a separate realm, since *we* are not a separate realm. The living world is made up of linked, interdependent, and interwoven alterities.

If we no longer think of ourselves as "humans" in the face of "nature," but as living beings among living beings, we no longer protect nature as something "alter" that is wild, or that is a fragile resource – we defend the community of living beings to which we ourselves belong, the community that has made us and that keeps us alive. To defend the living world is to shatter the false alternative that requires us to choose, a priori and in general, between nature and humans. (Local conflicts over priorities will, of course, still emerge, but they are already being configured differently.)

We need to think of humans as living beings who are part of the community of living beings, but with all their irreducible originality – as much in their capacity to care as in their capacity to destroy; as much in their unique capacity to plan as in their capacity to blind themselves.

To say that humans are members of the community of living beings does not mean that humans are part of nature, it does not mean returning to the fusional monism in an all-encompassing Nature, because, after all, there is nothing to protect if *everything* is nature. It means something else entirely. It means defending our multispecies milieus of life.

The protection of nature is not a hobby for greens, it is the name of our relation to the world. But it is not nature, in the old sense, and it is not protection. Protecting nature means defending the dynamic interweaving of our kin, the living beings that have made us, that constitute us, and that keep us alive from moment to moment. Defending the interweaving, apart from what would be useful or useless for us. Accepting the fact that we are dealing with trajectories in transformation and defending the potentials for evolution and the arrangements for equilibration in those trajectories.

The living world is not a domain of objects, it is the community of the world to which we belong: it can never be outside us; it is we who are inside. We are the living world defending itself.

## *Who makes the world inhabitable?*

What is at stake is a recharacterization of the nature of our interdependence with the living world. It is commonly acknowledged today that humans depend on the biosphere. But this recent awareness, although salutary, is limited, because it is incapable of getting beyond dualism. To be sure, humans are no longer considered as extracted from the milieu, but this milieu remains external, coded as

## Conclusion: The Living World Defends Itself

a reserve of resources that must now be managed in a more durable way, for our survival. But real interdependence goes deeper than this. It could be formulated through a simple question: who makes the world that makes us live? Just look around. Suppose you are in a city. You live every day in the midst of a milieu entirely made by human hands: buildings, streets, vehicles, lights, transformed foodstuffs. Everything leads to the belief that it is technological genius, human intelligence, and human work that make the planet inhabitable. This has some basis in fact, but the essential truth lies elsewhere. This everyday experience of city life conceals the fact that we ourselves are not responsible for the earth's inhabitability: the other living beings are the ones that make the earth inhabitable. Cities create a kind of obliviousness, an invisibilization, a sleight of hand, that makes us transpose the experience of this world created by humans to the whole of the planet. Let us try a thought experiment. If we suddenly remove, via our imaginations, everything the other life forms do to make the world inhabitable, humans die three times over: in 96 seconds, owing to the lack of oxygen; in 3 days, owing to the lack of potable water; in 3 weeks, owing to the lack of food; what ensues is an earth that is uninhabitable at all levels. All the clothing that protects us is made from living beings or hydrocarbons, which are derived from fossilized life forms; all mechanically powered means of transportation burn energy derived from ancient living beings.

What makes the world that keeps us alive? Plants and phytoplanktons, bacteria and viruses, earthworms, fauna in the soil and pollinators, and in general the dynamics of life ensured by the communities of living beings: by resilient and healthy ecosystems.[8] Who among us has integrated into their personal, everyday cosmology the fact that the formation of clouds, those that ensure the cycle of water circulation that keeps us alive, is maintained in many cases by populations of extremophile bacteria?⁹ Every hailstone from a storm cloud harbors a bacterial life almost as rich as that of a river. These bacteria favor the formation of hydrometeors that ensure precipitation. Storm clouds are among the most extreme habitats of our world, and yet they are inhabited, and "produced," by life forms whose power to guarantee the cycle of precipitation — to bring water back to the rivers, to the phreatic aquifers and water tables — escapes us even as it brings us our drinking water.[10]

We living humans can help to make the world a little *more* inhabitable, in the sense of making it more comfortable for ourselves, but we cannot *make* a world that is inhabitable for all forms of life. Everything that makes a world, starting from a ball of accreted matter that we call a planet, is an effect of living beings: we inhabit the effects of the life of the others. We are not the artisans responsible for the inhabitability; the other living beings play that role. It is their interweaving that ensures the continuing inhabitability of the world: symbiotic mushrooms and photosynthetic organisms and guilds of herbivores, hymenoptera and birds, collembola in the ground, viruses and bacteria, these are the ones that keep the world going and make it capable of harboring us, feeding us, healing us — us and all the others.[11]

People often speak of a contemporary "disconnect" from nature, especially on the part of city-dwellers, who are seen as "cut off" from nature. This hackneyed formula may find a solid meaning here: the disconnect is not a matter of geographic distance, nor is it a matter of love; it is not that urbanites do not love "nature," or love it less than those who live in close contact with what is called nature. It is that, inasmuch as they live in places marked from floor to ceiling by human hands, they are more likely to forget how our world holds together, this world woven of 4 billion years of evolution of living beings that

## Conclusion: The Living World Defends Itself

are everywhere and that keep it going. To be disconnected from nature, if this formula has a meaning, means above all to be distanced from sensitivity to *who* has actually made our world: the pollinators that literally make spring happen; the fauna in the soil that allow harvests to come back every year – and only secondarily the work of farmers; all the wild plants and phytoplankton that contribute to the production of oxygen, and so on. If there is something to be said about the relation to nature produced by the sociology of a massive rural exodus, in my view, it is this: it is amnesia concerning who *produces* and ensures the inhabitability of the world. And this inhabitability is not *for* us, nor is it reserved *to* us: it is for the fabric of life of which we are members, since there is only one world.

The shift of food production to sites far from dwelling places, the urban translation of food into merchandise separated from the soil and exchangeable for conventional currency earned by the work of individuals, the devaluation of traditional farmers' conceptions of the world – all these arrangements have helped to make the milieus that support us invisible in their roles as donors and world-menders. But we need to look deeper still: at issue is our relation not simply with food but with the entire living cosmos, inasmuch as that cosmos ensures, without awareness or intent, that the earth remains inhabitable for us. Because we have coevolved with it and in it and thus have adjusted to it, because we have been made by it and thus have been spontaneously fulfilled by it – since all the versions of ourselves that were not adjusted to this world have disappeared. We are so intrinsically of this world, so dependent on this world, so unmistakably made by this world for this world, that we begin finally to glimpse the folly of the gnostic formula "we are not of this world," and to grasp the extent to which it has founded one part of the modern relation to the milieus that are coded as "nature," as distinct from "human."

## Rekindling Life

One of the challenges of the cultural battle to restore importance to the living world consists, consequently, not in "getting closer" to nature but in recalling who are the most important workers when it comes to making the world inhabitable. We are not – we shall never be – those workers.

Humans are latecomers in the history of life: we are the ambitious nouveau riche players who have conquered all the managerial positions in the biosphere and have put to work all the others – the pollinators for truck farming, the soil fauna for agricultural fertility, the forest and marine planktons for the production of oxygen. Now, these nonhuman workers are fairly silent, not very politicized, not very demanding; they really do not have a culture of struggle and social protest. They are not the most eloquent advocates for their own importance. They just quietly perform their daily miracles as soil fauna that keep the entire landscape alive, as alluvial forests that filter water, as pollinators that fabricate springtime; they make no militant claims on our attention.

So certain humans have to betray their camp, the management side, and become union delegates advocating for living beings. "It is difficult to be a traitor," said Gilles Deleuze; "it is to create. One has to lose one's identity, one's face, in it."[12] There is no doubt about it: one has to lose one's old face as a human detached from kinship and interweaving before becoming a different human.

Can there be union representation for the living world? This strange and necessary job description was invented by James Lovelock. "Our union represents the bacteria, the fungi, and the slime moulds as well as the *nouveau riche* fish, birds, and animals and the landed establishment of noble trees and lesser plants. Indeed all living things are members of our union, and they are angry at the diabolical liberties taken with their planet and their lives by people."[13] This is another

## Conclusion: The Living World Defends Itself

way of conceptualizing the anthropological difference: the distinctive feature of human life (since there is such a thing, as there is for every form of life) is not that we are sovereigns and owners of the Creation, nor that we are stewards of planetary resources – we are workers in the field of common inhabitability, a little more diplomatic than some others, and combative against injustice. "We are not managers or masters of the earth, we are just shop stewards, workers chosen, because of our intelligence, as representatives for the others, the rest of the life forms of our planet."[14]

The difference between a bee and a human, here, is that the human is a slightly more noisily demanding worker than a bee, even if the human is a much less decisive factor than the pollinator in the fabric of the common world shared by these life forms.

Lovelock's metaphor is powerful because it displaces the focus: from the standpoint of human cognitive faculties, yes, humans are incomparable – this is a truism; but from the standpoint of Gaia – that is, considered in terms of the contribution to the *inhabitability of the earth*, in terms of how the vital conditions of a common world are produced – we are infinitely less important workers than phytoplanktons or earthworms. Here is where the decisive gap plays out: from the standpoint of real production (not the production imagined by the modern political economics of improvement), the real producers of biomass, abundance, and security are not what we think – they are not the Moderns, the arrangers, the improvers; they are not we humans. They are the nonhuman living beings. And the drama in this affair is that we really see this only when we kill them: when we systematically weaken their ecological functioning – which we had already naturalized, banalized, transformed into a fixed asset.

The union delegate is a good metaphor for the problem that concerns us, for it is inclusive: union delegates protect all workers, themselves included – they are not on the outside. The human world is outside of what is protected when we "protect nature"; it is inside when what we are defending is the community of living beings, a community to which humans belong. It is thus not "nature" that is targeted, because protecting "nature" excludes ourselves, replays the dualism that we are seeking to defuse, in a form that is either antihumanist (as manifested in the misanthropy of cults of wildness) or else hyperhumanist (as manifested in our pride in being the only moral species, the one that saves the others).

The philosophical move that needs to be made thus has a two-step trigger. The first step is, of course, to repopulate the world, to shift living beings from the status of décor or exploitable resource to the status of cohabitants, making it clear that they inhabit the same world as ourselves. But then – and this is even more dizzying – comes the challenge of the second step: to incorporate into our everyday metaphysics and our most systematic practices the fact that these other living beings are not simply inhabitants – they actually *create the inhabitability of the world* for *all living beings*, ourselves included. Defending life, defending the interspecies milieus of life, is a way of defending the inhabitability of the world – for us and for all the others, because we are all interwoven. Because it is the same world, and because we are part of the same fabric.

## *Retaking charge*

Reconceiving the "protection of nature" as a defense of the fabric of the living world has practical implications. In effect, it means freeing this endeavor from its confiscation by conservation experts, on the one hand, and by nation-states, on the other.

## Conclusion: The Living World Defends Itself

Civil society strongly feels the need for reappropriation, as we can see from the multiplication of individual and collective initiatives to buy back forests, create zones to be defended, fight against pesticides, and so on. The recourse to private property in various initiatives is, in this connection, a symptom: it is a lever that can be manipulated directly – by those who own land, by groups that organize to acquire land. In the French context, let us recall that nearly three-quarters of our forest lands are in private hands; these are the ones that need to be mobilized. In a parallel move, in the French context, the National Forestry Bureau (ONF) needs to be defended as a robust and well-endowed public agency, at the service of forests and of a sustainable production of lumber that does not violate forest dynamics. This does not mean that these initiatives bow to private property and its "world." Turning these milieus into commons, recuperating them without the artifice of property, is indeed the philosophical undergirding that animates such initiatives, but we still lack the functional legal tools to get around the issue of ownership and activate the commons collectively. The recourse to ownership responds to the feeling of urgency in a given situation; this is, in fact, how the initiative of the Réserves de Vie Sauvage saw the light of day. Inhabiting a zone to be defended, establishing a commons, becoming collective owners – these are three facets, each with its own strengths and weaknesses, of a powerful broader movement that entails democratization and citizen reappropriation of care for milieus, a movement that reclaims the defense of the fabric of life.

Still, this reclaiming by citizens does not mean that national governments can disengage – on the contrary, they must be alert to this lack of satisfaction with their own actions and rise to the occasion. In the same way, the collective and individual reclamations that work through non-governmental initiatives must not imply disengagement on the part of citizens toward the issue of public policy. The struggle occurs on multiple fronts. Starting a campaign, exerting pressure democratically to improve environmental law in terms of its capacity to punish environmental crimes, is of course one of the major axes of this reclaiming. At the same time, we must no longer allow care for the milieus of life to be confiscated: we *are* the life that is defending itself.

Indeed, leaving the protection of nature to conservation experts and national governments alone has some significant disadvantages: experts are sometimes prey to "managitis". As a local technique, this is sometimes appropriate, but as a monopolistic drive it has become a problematic mode of management. Thus, defenders of free evolution can no longer count on conservators of natural spaces and their expertise alone to protect milieus. For its part, the French state is compelled to enter into multi-user negotiations with actors (hunters, for example, or union representatives of extractivist agriculture) with whom private citizens may no longer want to negotiate over certain milieus. Let us recall in passing that actors such as these do not negotiate their uses when they are owners. For citizens to take charge of the defense of the fabric of life is thus the beginning of a quite significant development, since it means reappropriating a task that had been under the exclusive control of national governments, enmeshed in serious collusion with lobbies of destruction. When citizens retake control over the defense of their milieus, this gives much greater traction to protective measures. If tens of thousands of landowners mobilize, if myriads of citizens struggle to protect patches of the milieu more or less everywhere – fragments of forests, segments of rivers, wilderness areas, agricultural zones around their homes – against destructive uses of pesticides, then care for milieus takes on a new visage: citizens assume anew that they are woven

## Conclusion: The Living World Defends Itself

into the milieu; they defend the interdependencies between themselves and the multispecies landscape. This can change the situation on a broad scale, even as it resolves the fundamental aporia of the old protection of nature: reconciling inhabitants and their living milieu.

Nevertheless, my intention here is not to set the various modes of defending the fabric of life in opposition to one another: all modes of care are allies when they are meant to rekindle the embers of life (national parks, conservation bodies, initiatives by citizens or associations). The question is how to articulate them. I do not mean, either, to reject the role of the state in these matters, on principle – that would be immature. Public policies and the forging of laws have to remain spearheads of the defense of milieus, and there is no question of ceasing to militate in favor of electing representatives capable of advancing the cause. What I am criticizing here is the *monopoly* on the part of the state and of experts in the defense of milieus; this monopoly is already challenged when mayors oppose the state on the issue of pesticides, for example, or when people in a zone to be defended squat in a marsh. Paradoxically, these two types of resistance are two faces of the same basic movement: reclaiming the defense of our multispecies milieus.

### *The cultural battle*

Still, so that everyone can take part in defending the fabric of life, as living beings woven together with others, we have to face up to a cultural battle concerning our representation of ourselves. The problem can be formulated as a paradox. The thesis of our interdependency and our common kinship with the rest of the living world is well established in our cultural space, yet the living world is still not central in the field of our collective attention – in the political field of what concerns us first and foremost as a society, or even in contemporary ecological thought. The status of a human as one form of living being among others is not yet part of our cultural conception of ourselves.

The formulation of the paradox can be sharpened even further: in our tradition, the evolutionary and ecological forces of the living world, which have fabricated us body and mind, which give us all our powers of joy, thought, feeling, and connectedness, which keep us alive every day, which ensure (without intention) the earth's inhabitability – these forces are devalued, hidden, marginalized, and people who show them even minimal gratitude are mocked for their devotion to flowers and beasts. Here is where the immense misunderstanding plays out: by marking those who demonstrate consideration for the living world with the stigmatizing label "nature lovers," dismissing them as depoliticized dreamers or naïve utopians, our tradition has operated the most effective hostile takeover imaginable for turning the human collective away from its world, its family, its origins, its satisfying interweaving – that is, all the rest of life.

Here is the great oversight and the great misunderstanding of the dominant modern philosophical anthropologies, especially those of the West. Life is conceptualized as a domain encompassing the objects of the world, among other domains, but not as our own most profound identity. We are living beings among other living beings, shaped and irrigated with life every day by the dynamics of the living world. Claude Lévi-Strauss formulated the problem brilliantly, arguing that the nature vs. human dichotomy is a defensive construct developed by our culture to mask our "original association [*connivence*] with the other forms of life."[15] We need to blow up this defensive construct, dynamite it so we can feel ourselves to be living beings interwoven

## Conclusion: The Living World Defends Itself

with others, *before* we feel ourselves to be human, and modern. The old dualism in effect hides our real identity, our vital *connivence*. It is, of course, not the product of culture in general – that is, of all culture – but of one specific culture. Exiting from dualism thus does not imply doing without culture and returning to nature (in a relapse into the chronic dualism); it implies creating a *different* culture that does not mask our initial association with the other manifestations of life, but instead celebrates it. This is the sense in which we are dealing with a real cultural battle to be waged around the question of the living world.

Here is the paradox: to contribute to a future that ensures care for *humans*, we have to transform our philosophical anthropology into one in which we do not identify ourselves *first and foremost* as humans. For the good of humans, since this is obviously a central stake in the crises to come, we must think of ourselves first of all as living beings. In the same way that, for the good of every skin color, we have had to think of ourselves as human before thinking of ourselves as white or black: we have had to choose a larger set in order to recognize ourselves in a way that protects the independent plurality of that set. But the inclusion here is not the same as inclusion where the issue of racism is concerned: it is not a matter of setting up living beings in a category where each one is a person, in an egalitarian framework. With the bacteria on my skin, the fauna in the ground, all the actors in the dynamics of life, the whole question of equality does not pertain: we are no more equal than we are unequal – we are interdependent. And it is the question of adjusted consideration in a myriad of relations (where one lives by eating one another, by disturbing one another), at every scale, from the ecosystem to populations to individuals, that becomes the relevant site of controversy today, in my view.

## Rekindling Life

To respond to the devaluation and obfuscation of the importance of the living world on which we are grounded, calling on a collective and democratic inquiry to invent adjusted consideration toward life forms is entirely different from calling for a cult. There would be no more badly adjusted, more disastrous solution than a cult of the living world in the sense of a religious cult aiming toward a reinstituted transcendency. The humanist approach entails a visceral repugnance toward anything that, seen from close up or from a distance, resembles a transcendency on which humans would depend. This is partly justified, partly toxic. From the humanist standpoint, we are duty-bound not to admire or show consideration for anything but the human, for fear of falling back into the cult of external idols (Nature, God, and so on), those transcendencies from which humanism offers precisely to liberate us. To admire nothing but itself as an emancipatory power – here is a founding affect of modernity. It was a liberating habit for a time, for certain struggles, but, pushed to the limit and made an absolute principle, it is a form of alienation. It is not a question here of calling for a cult of nature – there is no nature; the challenge is to show consideration, based on good sense, for the powers of life within us and outside of us, in a way that implies no cult other than gratitude, trust, reciprocity, and *the tireless collective invention of adjusted consideration*. In other words, no cult other than a radical struggle against the economic and political forces that weaken and destroy those powers of life.

One example of those negative forces, as we have seen, is extractivist agriculture, which proceeds from the metaphysics of "improvement" as an absolute principle – one that is embodied in Europe in the Common Agricultural Policy. At bottom, this is a metaphysics of "production." Put in more general terms, it is obvious that a dominant visage of the enemy is embodied in contemporary capitalist

## Conclusion: The Living World Defends Itself

arrangements built on a principle of unlimited "progress," a cult of yields, an enthronement of growth as a primary indicator, a commitment to productivist extractivism, and a consecration of inequalities. But to put things this way is still to remain too abstract: we need to go into particulars, in the imbroglio of practices and institutions, if we are to identify the enemies of the living web with finesse and come up with truly effective levers to neutralize them.

In this battle, we are not a solitary species confronting the rest of the world packaged as "nature": we are not face to face, but side by side, with all the other living beings, confronting the theft of our common world. There are uncountable numbers of us ready to weave and defend the inhabitability of this world.

# Notes

## 1 Give Us a Lever and a Fulcrum

1 https://ipbes.net/sites/default/files/ipbes_7_10_add.1_en_1.pdf.
2 This is the celebrated "lever effect," which multiplies the effectiveness of the person using a lever. The invention can be used for better or for worse: lever effects are used in global finance in the form of leveraged loans destined to multiply profits for the lender, but to the detriment of the world economy, as we saw with the subprime crisis in 2008.
3 [Translator's note] The word *milieu* is defined in the Merriam-Webster online thesaurus as referring to "the place, time, and circumstances in which something occurs," especially "the physical and social surroundings of a person or group of persons" (https://merriam-webster.com/thesaurus.milieu). In this text, the term is used more broadly to encompass the physical and social surroundings of a living being or a group of living beings.

## 2 Anatomy of a Lever, A Case Study: Hearths of Free Evolution

1 The Association pour la Protection des Animaux Sauvages and the Forêts Sauvages association already manage six wildlife preserves spread throughout France. The idea of land acquisition was originally tied to that of free evolution in the thinking of the inexhaustible naturalist

Gilbert Cochet and the intellectual team that gravitated around Forêts Sauvages. I would like to take this opportunity to pay homage to the tireless, original, and powerful work of the founders of the Forêts Sauvages newsletter, and its contributors (P. Athanaze, B. Cochet, G. Cochet, J.-C. Génot, O. Gilg, C. Gravier, P. Lebreton, M. Michelot, J. Poirot, A. Schnitzler, C. Schwoehrer, J.-L. Sibille, L. Terraz, D. Vallauri). Since 2007, they have been plowing the furrows of free evolution, land control, and the philosophical stakes of the relation of humans to forests, in the invaluable journal *Naturalité, la lettre de Forêts Sauvages*, readily accessible on the Internet at www.forets-sauvages.fr/web/foretsau vages/100-naturalite-la-lettre-de-forets-sauvages.php.

2 The expression is from Marina Fischer-Kowalski and Thomas Macho, *Gesellschaftlicher Stoffwechsel und Kolonisierung von Natur: Ein Versuch in Sozialer Ökologie* (Amsterdam: G+B Verlag Fakultas, 1997). Free evolution constitutes a form of management that entails total withdrawal from any active development of a forest: it consists in eliminating the modes of intervention (forcings) that act on forest ecosystems (seeding, treatment, management, cutting) with the aim of freeing the dynamics and potentialities of the milieu. As we shall see, this management by withdrawal, used sparingly, is also the style of nonviolent silviculture, except that the latter maintains certain forms of management and action, such as tree removal.

3 The estimates vary according to the types of forest and the approaches adopted, but the general tendency is clear. On this point, Baptiste Regnery establishes positive correlations between biological diversity, the degree to which trees abound in microhabitats, and the age of the forest. His approach is interesting in that it provides an indicator available to foresters: they can evaluate the number and diversity of microhabitats in order to determine and enrich the diversity of life forms in a forest. See Baptiste Regnery, Yoan Paillet, Denis Couvet, and Christian

Kerbiriou, "Which Factors Influence the Occurrence and Density of Tree Microhabitats in Mediterranean Oak Forests?" *Forest Ecology and Management* 295 (May 1, 2013): 118–25; and Baptiste Regnery, Denis Couvet, Loren Kubarek, et al., "Tree Microhabitats as Indicators of Bird and Bat Communities in Mediterranean Forests," *Ecological Indicators* 34, no. 1 (November 2013): 221–30.

4 Emma Marris makes this point in *Rambunctious Garden: Saving Nature in a Post-Wild World* (New York: Bloomsbury, 2013).

5 Émilie Hache, *Ce à quoi nous tenons* (Paris: La Découverte, 2014).

6 In this connection, see the synthesis offered by Laurent Godet and Vincent Devictor concerning the relationship between policy and conservation; the authors analyze more than 13,000 articles – essentially the entire set of publications in the discipline published in the nine largest journals in biology and conservation between 2000 and 2015 ("What Conservation Does," *Trends in Ecology & Evolution* 33, no. 10 [October 2018]: 720–30). Godet concludes, for example, that "conciliation always privileges the economy over environmental interests. Protection of species and spaces suffers from these compromises: protected zones are not truly protected when tourism and agropastoral activities are present. Truly protected soils in integral biological reserves amount to only 0.02% of the territory of metropolitan France": see Usbek and Rica's interview with Laurent Godet, "La survie du monde vivant doit passer avant le développement économique." *Usbek et Rica*, September 10, 2018, https://usbeketrica.com/fr/article/la-survie-du-monde-vivant-doit-passer-avant-le-developpement-economique.

7 This is spectacularly apparent in the case of the red mud in the Calanques National Park in southern France, where no one seems capable of preventing industrialists from polluting a precious protected zone.

8 See Pierre Athanaze, "Des réserves de vie sauvage," *Naturalité, la lettre de Forêts Sauvages*, no. 13 (February 2014): 2.

9  See Alain Persuy, "Pour des forêts vivantes, résistance!!!" *Naturalité, la lettre de Forêts Sauvages*, no. 18 (December 2018): 8.
10 See Jean-Claude Génot, *La nature malade de la gestion* (Paris: Le Sang de la terre, 2008).
11 On this point, see François Sarrazin and Jane Lecomte, "Evolution in the Anthropocene," *Science* 351, no. 6276 (February 26, 2016): 922–3, for important ideas concerning an "evocentric" approach to the conservation of nature.
12 See Vaclav Smil, "Harvesting the Biosphere: The Human Impact," *Population and Development Review* 37, no. 4 (2011): 613–36, and Anthony D. Barnosky, "Megafauna Biomass Tradeoff as a Driver of Quaternary and Future Extinctions," *PNAS* 12, no. 5, suppl. 1 (2008): 11543–8. There are significant biases in these models of which the authors are aware. But I think these estimates are accurate for the most part. It is even probable that the biomass of wild fauna 10,000 years ago has been considerably under-estimated, although this is hard to demonstrate. One can also calculate measures in the absolute, rather than in proportionate terms. Moreover, the "biomass" metric is very partial, in both senses. To expand the analysis, we would need to multiply the modes of quantification, but the data for the other metrics are even more fragmented: the number of individuals, the zones of distribution, species-specific and intraspecific biodiversity, and so on. We would nevertheless need all this information to have a clear sense of the tragedy.
13 Karl Heinz Erb, T. Kastner, C. Plutzar, et al., "Unexpectedly Large Impact of Forest Management and Grazing on Global Vegetation Biomass," *Nature*, no. 553 (2018): 73–6. This is entirely compatible with the fabrication of a biomass of domestic herbivores and humans to the detriment of wild beings.
14 Gilbert Cochet, personal communication.

15 On this point, see Alexandre Robert, Colin Fontaine, Simon Veron, et al., "Fixism and Conservation Science," *Conservation Biology* 31, no. 4 (August 2017): 781–8, https://pubmed.ncbi.nlm.nih.gov/27943401.
16 Gilbert Cochet, personal communication.
17 This is documented for protected marine zones, which, when they are well defined and respected, increase the income of the local fishermen. In ecology, one speaks of "sources" or wellsprings to characterize zones where the population dynamics are on the rise. Clearly, the hearths of free evolution are wellsprings for species for which, by contrast, highly exploited milieus are bottomless pits. There are relevant controversies concerning habitats: for example, can a forest in free evolution claim to constitute a wellspring for birds whose usual habitats are not in forests? It cannot; but it can be a wellspring for more polyvalent species and for specific forest birds that, as has been shown, suffer from the practices of intensive agriculture – not to mention intensive forestry.
18 Gilbert Cochet formulates the link between engagement in favor of wild forests and struggle against climate change in an amusing equation: "One hectare of forest stores between 9 and 10 tons of $CO_2$ per year. Every French person expels on average 7.5 tons of $CO_2$ per year. If one Frenchman buys one hectare of forest in free evolution, he is 'carbon compensated.' But be careful, it isn't like buying indulgences to ensure that one will get to heaven (by plane). One has to maintain frugal and virtuous behavior tending toward a minimum expulsion of $CO_2$. And compensate for the rest."
19 See Georges Canguilhem, "The Normal and the Pathological," in *Knowledge of Life*, trans. Paola Marrati and Todd Meyers (New York: Fordham University Press, [1952] 2008), pp. 121–33.
20 On this point, see the crucial report edited by Gaëtan du Bus de Warnaffe and Sylvain Angerand on the future of forest management in the face of climate change: the authors show convincingly that a

scenario calling for leaving 25 percent of France's forested mountains in free evolution, favoring sustainable management of a continuous silviculture type, and preferring high-quality durable uses of wood products (carpentry, woodworking, and so on), is the best plan in the medium term, the one that will allow forests to play the most decisive role in carbon storage. See *Gestion forestière et changement climatique: Une nouvelle approche de la stratégie nationale d'atténuation*, January 2020, accessible on the Alternatives forestières website: www. alternativesforestieres.org/IMG/pdf/synthese-web-rapport-foret-climat-fern-canopee-at.pdf.

21 "Charte des réserves de vie sauvage." See the ASPAS site: https://aspas-reserves-vie-sauvage.org.

22 ASPAS has a slightly different legal status, under local laws, because its headquarters are in Alsace. But the principles are the same.

23 Open letter from Philippe Falbet to the Confédération paysanne de l'Ariège. Aspet (Haute-Garonne), May 17, 2019.

24 For a reflection of another sort on the link between the notion of "commons" and nonhumans, see an article by Lionel Maurel, accessible on the blog S. I. Lex: "Communs & non-humains (1ère partie): Oublier les 'ressources' pour ancrer les Communs dans une 'communauté biotique,'" January 10, 2019, https://scinfolex.com/2019/01/10/communs-non-humains-1ere-partie-oublier-les-ressources-pour-ancrer-les-communs-dans-une-communaute-biotique.

25 ASPAS has imagined an Internet site, moreover, that would generate a sort of title of ownership based on the gift, indicating the amount of land that the donation made it possible to give back.

26 Notre-Dame-des-Landes is a Zone to Be Defended (*Zone à défendre*, or ZAD) – that is, a terrain inhabited and defended by citizens mobilized to prevent the installation of a Big Useless Project (*Grand projet inutile*, or GPI). In the case of Notre-Dame-des-Landes, the GPI entailed

the construction of a new airport. That project was abandoned by the French government in 2018, after several years of mobilization and confrontations between Zadistes and official security forces.

27 This obligation applies for a period of 30 years, by contractual agreement between a landowner and a third party (the original measure called for non-exploitation in perpetuity, but that provision was not included in the 2016 law). Thanks to Sarah Vanuxem for drawing my attention to this measure, and for her generous reading and critique of this text.

28 Lionel Maurel, "La propriété privée au secours des forêts ou les paradoxes des nouveaux communs sylvestres," an article published August 19, 2019, accessible on the blog S. I. Lex: https://scinfolex.com/2019/08/19/la-propriete-privee-au-secours-des-forets-ou-les-paradoxes-des-nouveaux-communs-sylvestres.

## 3 The Embers of Life

1 By "life," I am referring here to the entire set of dynamics and processes of eco-evolution. Notre-Dame is a few centuries old, while a lineage of bees, bacteria, orangutans, or humans carries within it 3.8 billion years of transformations of life.

2 Sébastien Dutreuil recalls this magnificently in his reflection on the conception of life that underlies the Gaia hypothesis: see his article "Quelle est la nature de la Terre?" in Frédérique Aït-Touati and Emanuele Coccia, eds., *Le cri de Gaïa: Penser la Terre avec Bruno Latour* (Paris: La Découverte, 2021), pp. 17–65. He writes: "It is as though the scientific work on Gaia had been carried out on the basis of a conviction that was never questioned: the vital biotic forces exercised on the environment or the milieu know no limits and always exceed the physical, chemical, and geological forces . . . Certain passages [in

James Lovelock and Lynn Margulis] attest to that metaphysical intuition concerning the expansive character of life: 'Life tends to grow until the supply of energy or raw materials set [*sic*] a limit. Probably a planet either is lifeless or it teems with life. We suspect that on a planetary scale sparse life is an unstable state implying recent birth or imminent death'" (cited from James Lovelock and Lynn Margulis, "Biological Modulation of the Earth's Atmosphere," *Icarus* 21, no. 4 [1974]: 486).

3  It is an intuition of this adventure that Nietzsche was transcribing when he maintained that all generosity is an overabundance of force: "Life as a whole is *not* a state of crisis or hunger, but rather a richness, a luxuriance, even an absurd extravagance – where there is a struggle, there is a struggle for *power*" (Friedrich Nietzsche, *The Twilight of the Idols*, trans. Duncan Large [New York: Oxford University Press, 1998], ch. 9, section 14, "Anti-Darwin," p. 50).

4  Charles Darwin, *On the Various Contrivances by Which British and Foreign Orchids Are Fertilized by Insects, and on the Good Effects of Intercrossing* (London: J. Murray, 1862).

5  Some readers may be uneasy here, left with a vague feeling that the metaphor is not appropriate, imagining behind the word "fire" the Australian or Amazonian megafires. What a megafire, as a media archetype of life crisis, captures and triggers is the ancient trauma of the fire that burns down "my" house, the fire that destroys the world that I have made for myself and that will have to be rebuilt from scratch, stone by stone, out of the smoking ruins. This image of a megafire that destroys a "home" as the symbol of the anthropocenic destructions, so fitting in our minds, is in fact an unconscious trap that leaves its implications unseen. For here we are talking about another world (the real one), the common house: "we" did not build it, it is not a static construction subject to entropy, and "we" cannot rebuild it. It built itself, and it has

made us, through its ecological and evolutionary dynamics, through the history of life on earth. The problem of how to defend it has to be formulated differently. Fire must be fought with fire: we need to fight the figure of destructive fire with that of creative and regenerative fire. For it is the latter that restores to life on earth its own powers, and designs a better-adjusted relation between ourselves and that life, far from the Promethean pride and shame of a humanity that burns its creation, forgetting that it has not built this hearth. Forgetting, above all, that this hearth is first and foremost a power of autonomous reconstruction, if we allow it, or restore to it, the right conditions. The metaphor of life as fire reconnects us with a more just, better adjusted, more activatable, more highly developed relation with its target: life itself. It is in fact to fight against megafires that we need to rekindle the embers of life.

6 *Naturalité, la lettre de Forêts Sauvages*, no. 4 (April 2008): 4.
7 On this topic, read the excellent article by Raphaël Larrère, "Le réparateur, l'ingénieur ou le thérapeute?" *Sciences, eaux & territoires* 3, no. 24 (2017): 16–19.
8 On the rise in populations of marine species, including humpback whales, see Carlos M. Duarte, Susana Agusti, Edward Barbier, et al., "Rebuilding Marine Life," *Nature*, no. 580 (2020): 39–51. This important article shows that we can envision rediscovering the prodigality, health, and population numbers of the past in the world's oceans within 30 years, if we connect sustainable fishing, the fight against pollution and destruction of marine habitats, and the fight against climate change. Successes in conservation have shown the resilience of the seas. What is fascinating is that the very article that isolates the pyric power of marine ecosystems to reconstitute themselves *on their own*, once our destructive practices are stopped, is nevertheless titled "*Rebuilding* Marine Life." The imaginary Promethean realm of fabrication, restoration, reconstruction of something that in fact cannot be rebuilt by

us but that rebuilds itself, is a realm that spares no one, not even the conservationists who are the most enlightened about the autonomous powers of life.

9  As we shall see later on, it is a question of valorizing "direct negative action" to the detriment of "direct positive action": inserting ourselves into the strength of things, slightly inflecting the ecological dynamics on a single point to invigorate them, to let them express their full power. In contrast, direct positive action implies a relation to life that first requires making it dependent in order to make it manipulable, and thus induces, in a second phase, a paternalistic need to direct and protect it.

10 Some may be shocked by the metaphor of fire, since it has been used in our history mainly as an image of destruction (for example in the parable of the hummingbird fighting the forest fire one drop at a time, "doing the best he can"), rather than as an image of life as it spreads. But, quite apart from metaphors, the ecologic reality of fire is something entirely different from spectacular, definitive destruction: in the great North American forests, for example, although fire constitutes a disturbance (it destroys biomass), it is a precious auxiliary helping to regenerate milieus; many species have evolved to welcome it and to profit from its effects, like lodgepole pine trees and blue jays.

11 Certain central properties of fire are not transposable in a way that pertains to conceptualizing life: life does not generate structures of high entropy, does not possess an intrinsic power of destruction. The analogy with fire has its limits, for it leaves out certain central properties of life: fire does not diversify itself, does not invent forms whose continuity and inventivity stem from the fact that they are made of memory in creative dialogue with the milieu and with other forms – the fact that they are covariant memories that are interpreted as something like musical scores by the milieu.

12  See J.-H. Rosny, *The Quest for Fire: A Novel of Prehistoric Times*, translated from the French (New York: Pantheon Books, [1911] 1967). A French film based on the novel, directed by Jean-Jacques Annaud, was released in English in 1981.
13  See Raj Patel and Jason Moore, *A History of the World in Seven Cheap Things: A Guide to Capitalism, Nature, and the Future of the Planet* (Oakland: University of California Press, 2017).
14  Human peoples, in their cultural diversity, have explored a zone of possible ways of life, but that zone did not exist a priori, it has been opened up by exploration itself. The whole challenge is to explore it while maintaining the possibility of future explorations: by cherishing openness to the possible, and thus the durability of the conditions for life, the interweavings that maintain us in life all together. Sustainability is thus a necessary condition for the human adventure.
15  Philippe Descola, *Beyond Nature and Culture*, trans. Janet Lloyd (University of Chicago Press, 2013), p. 401.
16  The paradox according to which the defense of life offers protection to something more powerful than ourselves makes it possible to respond to those who criticize the very idea of "protection" of nature as the last colonialism, as a practice whose built-in ideology implies a downgrading of what is being protected. Speaking of "defending the embers of life" allows us to sidestep some of the problematic legacies of the idea of "protecting nature."
17  This is a technical rather than a moral argument against geo-engineering and bio-engineering as substitute "ecological services." This regenerative capacity is spectacularly visible on the old Chernobyl site, where prodigality in wildlife is well documented today. I should add, to do justice to technology, that if technology is not relevant when it claims to be a substitute for the powers that surpass us, it is useful and necessary as ecological engineering if its goal is to facilitate the removal

of constraints, or the activation of life's own powers (for example by reintroducing key species).

18  See Nurit Bird-David, "The Giving Environment," *Current Anthropology* 31, no. 2 (April 1990): 189–96.
19  In this connection, see the troubling book, with its sometimes questionable conclusions, by Chris Thomas: *Inheritors of the Earth: How Nature Is Thriving in an Age of Extinction* (London: Penguin, 2017).
20  See Gilbert Cochet and Stéphane Durand, *Ré-ensauvageons la France: Plaidoyer pour une nature sauvage et libre* (Arles: Actes Sud, 2018).
21  I have developed the idea of adjusted consideration more explicitly in Baptiste Morizot, *Ways of Being Alive*, trans. Andrew Brown (Cambridge: Polity, 2022), pp. 232–9: "Our era of systemic ecological crisis is a time in which our relations with animals, plants, and environments are at stake. These relationships must be reinvented . . . In animist cultures, the invariant of these relationships, what characterizes them all, is not abstract egalitarianism, but rather the fact that they always require consideration . . . the essence of the modern relationship invented by those who invented the recent idea of 'Nature' is the uselessness of consideration for living beings and non-humans: such consideration is irrational . . . What is at stake today is that our relations with the living world, with pollinating bees, ancient forests, farm animals, soil microfauna, and all the rest need to be reinvented: it is this consideration that we need to rethink . . . The choice of words is important here: 'consideration' doesn't sound like much, but it's actually an elaborate concept, a virtuoso at ducking and diving; it's a concept that slips in and out of the modern dualist division between our moral relation towards people (ends in themselves) and our instrumental relation to everything else (means for ends in themselves) . . . I want to redefine . . . the interspecies diplomacy of interdependences as a 'theory and practice of adjusted consideration.' The consideration that needs to be invented is 'adjusted'

and not 'fair,' precisely because the beings involved are beings whose powers, in truth, are unknown: their definitive moral status (person, dignity, end in itself, means, pure matter) is not available to us; we must constantly adjust and readjust our consideration for them depending on the answers they give us, on their ways of reacting, of altering our actions and sending them back to us in a different form."

22 PRELE is a regional program of spaces in free evolution; FRENE stands for Rhônalpine forests in natural evolution; RAF is a network for forest alternatives.
23 On this point, see the initiative of the Forêt Vivante en Haute-Savoie association: www.alternativesforestieres.org/Groupement-Foret-Viv ante-Haute-Savoie.
24 See Morizot, *Ways of Being Alive*, trans. Andrew Brown (Cambridge: Polity, 2022).

## 4 Realigning Alliances

1 The Confédération paysanne is a French agricultural union, situated on the left, that defends traditional small-scale farming, in opposition to France's largest agricultural union, the Fédération Nationale des Syndicats d'Exploitants Agricoles). The FNSEA, positioned on the right, defends conventional industrial agriculture.
2 See Pascale Laussel, Marjolaine Boitard, and Gaëtan du Bus de Warnaffe, *Agir ensemble en forêt: Guide pratique, juridique et humain* (Paris: Charles-Léopold Mayer, 2018).
3 This is not the place to write the history of the conception, but figures and premises for it can be found in conjunctures as remote as the ideology of "enhancing value" used to justify the system of *encomienda* in Mexico (appropriation of *terra nullius*, "nobody's land," large plantations, forced labor, evangelization of the Amerindians). While its

theoretical formalization came about in the seventeenth century, this unconscious conception had already structured the practices of the first colonizers in the Americas from the early sixteenth century on. But for a deeper genealogy, to be produced by better historians than I, we would have to go back to the great land clearings of the Middle Ages, which were often directed by local bishops and were thus given a reinforcing religious justification. The bishops were pioneers who founded rural or urban communities during the period when Europe was being repopulated after the fall of the Roman Empire. And the matrix for these foundations was "clearing" land to enhance its value in a durable way, to make it inhabitable. On these points, see Serge Gruzinski, *The Mestizo Mind: The Intellectual Dynamics of Colonization and Globalization*, trans. Deke Dunsinberre (New York: Routledge, 2002); Jacques Le Goff, "Ecclesiastical Culture and Folklore in the Middle Ages: Saint Marcellus of Paris and the Dragon," in *Time, Work & Culture in the Middle Ages*, trans. Arthur Goldhammer (Chicago, IL: University of Chicago Press, 1980), pp. 159–88; and also Jacques Le Goff (with Emmanuel Le Roy Ladurie), "Mélusine maternelle et défricheuse," *Annales. Économies, sociétés, civilisations* 25, nos. 3–4 (1971): 587–622.

4 On this concept, see Paul Warde, "The Idea of Improvement, c. 1520–1700," in Richard W. Hoyle, ed., *Custom, Improvement and the Landscape in Early Modern Britain* (Farnham and Burlington: Ashgate, 2011), pp. 127–48; Neal Wood, *John Locke and Agrarian Capitalism* (Berkeley: University of California Press, 1984); and, especially, Richard Drayton, *Nature's Government* (New Haven, CT: Yale University Press, 2000).

Pierre Charbonnier, particularly in *Abondance et liberté: Une histoire environnementale des idées politiques* (Paris: La Découverte, 2020), brings to mind the destiny of the term "improvement." Elsewhere, Charbonnier writes that "what makes the earth the prototypical object of economic and legal appropriation (and these amount to the same thing, since it is labor that gives the right to appropriate) is the fact that its initial state consists in a virtual support for productivity, fulfilled only by human labor, scientific knowledge, and technology. . . . Improving a plot of land, making it profitable by fencing it off and cultivating it, is thus the central operation of the political affordances of the earth at the time Locke was writing" (personal communication).

5 On this point, see a talk by Émilie Hache in which she presents her breathtaking critique of Judeo-Christian theology, by way of which we can be said to have "passed from a world in which we honored the mysteries of generation to a world in which we honor the mysteries of Creation": www.youtube.com/watch?v=9f-OyHRMMp-w&feature=youtu.be, 2020. The idea of the Creation in fact constitutes one historical matrix of the idea of "production": these two notions have in common the overvaluation of the agency of the author and the devaluation of that of matter, as distinct from generation, in which things themselves ensure their own reproduction.

6 Callicott describes "improvement" in these terms: "It seems clearly the intent of God that man be master and nature slave, since not only is man given dominion over the earth but he is expressly enjoined to subdue (Hebrew: *kabas*, 'stamp down') the earth – as if nature were created unruly and were in need of breaking to become complete" (J. Baird Callicott, *Earth's Insights: A Multicultural Survey of Ecological Ethics from the Mediterranean Basin to the Australian Outback* [Berkeley: University of California Press, 1994], p. 15).

7 This point stands out clearly when we know that improvement was also a colonial means "for distinguishing 'industrial and rational' men from the others, namely, the Amerindians: it is through this means, as has already been made clear, that indigenous societies are excluded from legitimate legal relations with land, since they are only hunter-gatherers"

(Pierre Charbonnier, personal communication). "Improving" amounts to taking charge of a particular local, historical, and ideological milieu and setting it up as an exclusive monopoly defining "good" use of the earth, thereby authorizing the exclusion of all other uses of a milieu, uses that do not "produce."

8 Unless we postulate that a god made us in his image and made the world for us, so that we could be its stewards. It boils down, ultimately, to a controversy in philosophical anthropology that in turn implies political cleavages.

9 Philippe Descola, *Beyond Nature and Culture*, trans. Janet Lloyd (University of Chicago Press, 2013), p. 326.

10 This thought experiment offers a new prism for understanding the stewardship model of environmental ethics, which is omnipresent in the Anglo-Saxon cultural sphere. In its unformulated premises, we can see how the "otherizing" of nature has been made invisible. For to postulate that it is now necessary to manage wild nature as good stewards, for its own benefit (operation 2), one must first have deemed it "other" (operation 1). Stewardship postulates, as an unquestioned fact, that nature needs humans to take care of it, and thus that to assume our role as steward is to engage in an act of humanity toward nature. But to postulate oneself as steward, one must first have made the fundamental gesture, both representational and technological, of heteronomizing the living world. We then forget that we have done so, and we set stewardship up as an ethics, even as we have already depreciated in advance what we mean to protect.

11 See Raj Patel and Jason Moore, *A History of the World in Seven Cheap Things: A Guide to Capitalism, Nature, and the Future of the Planet* (Oakland: University of California Press, 2017).

12 This is made explicit in Carolyn Merchant, *The Death of Nature: Women, Ecology and the Scientific Revolution* (San Francisco: HarperOne, 1990).

13 We should note that the phenomenon conceals an infernal flight forward: amelioration of the yield in consumable biomass begins as a struggle against famine, but it conceals the fact that the human species is a population of mammals like the others in certain respects. If the resource grows, the population proliferates, and so we find ourselves again in a headlong race, in a situation of lack and famine, and the self-fulfilling prophecy proves itself: we are committed to improving nature indefinitely in order to have enough resources for all, and to ensure the abundance and safety of the food supply.

14 Patel and Moore, *History of the World*, p. 47.

15 Calling them "ecosystemic services" can sometimes make sense politically, when it is a question of militating for a different common agriculture policy, for example, for this makes it possible to privilege through subsidies the exploiters who maintain and defend these dynamics, as opposed to the current perverse subsidies; in most other contexts, however, the formula is toxic. On this point, see the activities of the inter-organizational French platform Pour une autre PAC (For Another Common Agriculture Policy): www.pouruneautrepac.eu.

16 We can experience this for ourselves. Before I understood how forests worked, I felt uneasy when I was traversing an unmanaged forested area in the mountains: disturbed by the chaos of forms, the absence of paths, the dead wood on the ground, the dense thickets. I experienced an aesthetic uneasiness – anguish at being excluded. This still happens to me from time to time, especially when I am tired or absorbed in human preoccupations. It is fascinating to see how a constructed cosmology becomes inscribed in our most spontaneous, most instinctive perceptions. The challenge is to learn, little by little, that these landscapes, if they are not spontaneously welcoming to us, make sense in the ecology of the forest, from the forest's point of view: the dead wood fallen anarchically on the ground is biomass restored to the decomposers that

will supply life to the soils of tomorrow. For it is in these dynamics that forest-dwelling species have coevolved together.

17 [Translator's note] The term 'input' (French *intrant*) is used in agriculture and forestry to refer to any organic or inorganic product added to the soil or crop in a given milieu: fertilizers, herbicides, pesticides, and so on.

18 This is a key point: we must recall the mechanistic dimension of the problem in order to get away from a moralizing critique of the farmers themselves. This has been abundantly shown by others, but I can sketch out certain of these mechanisms here: competition that is virtually global; multiplication of intermediaries between producer and consumer, which makes prices go up at the end of the chain and go down at the start of the chain owing to the pressure of the intermediaries and distributors in the power relations – and this effect is proportional to the length of the long-distance supply chains.

19 See Jonathan Safran Foer, *We Are the Weather: Saving the Planet Begins at Breakfast* (New York: Farrar, Straus and Giroux, 2019).

20 Here I am again applying a methodological element that recurs frequently in this investigation: it consists in applying the lessons of tracking to the practices in question and to the language used by practitioners. It is a matter of being attentive to potential discrete elements, to details that may reveal underlying mentalities – that is, invisible metaphysical structures. The challenge is then to substitute other ontological maps for these often dualist legacies: maps backed up by other foundational intuitions that are already present among practitioners, for example farmers' confidence in the art of plants to grow; maps of experience endowed with a more viable practical aim, offering political potentials better adjusted to the necessary cohabitation with the rest of the living world.

21 This formulation was used by Vere Gordon Childe to characterize the "Neolithic revolution"; it has become a fixed element of the modern

understanding of the process of Neolithization. See Vere Gordon Childe, *Man Makes Himself* (New York: Oxford University Press, 1939). Catherine Perlès offers a good analysis of the phrase "from predation to production" in "Pourquoi le Néolithique? Analyse des théories, évolution des perspectives," in Jean-Pierre Poulain, ed., *L'homme, le mangeur, l'animal, qui nourrit l'autre?* (Paris: Observatoire Cidil des habitudes alimentaires, coll. Cahiers de l'OCHA, no. 12, 2007), pp. 16–29.

22 The prehistorian Jacques Cauvin has said the same thing differently: we had been "spectators of the spontaneous reproduction of living beings"; we then passed into "active production" – thus, from natural reproduction to production. Strange magic once again: do plants and animals no longer reproduce themselves? See Jacques Cauvin, *The Birth of the Gods and the Origins of Agriculture*, trans. Trevor Watkins (Cambridge University Press, [1997] 2000).

23 As the Latin *pro-ducere*, "to produce" also means to cause to appear, or to make manifest. It is clear that, in the origin of the word in French or English, there is no distinction between production by nature and production by human minds or technologies. This is evident in René Descartes's words: "Nor can we suppose that several causes may have concurred in my production" (*Meditations on First Philosophy*, in *The Philosophical Works of Descartes*, trans. Elizabeth S. Haldane [Cambridge University Press, (1641) 1911]).

24 Descola, *Beyond Nature and Culture*, p. 325.

25 Charles Darwin, *The Origin of Species* (New York: P. F. Collier & Son, [1859] 1909), p. 93.

26 Let us recall here, in passing, a slogan adopted by the Grenelle de l'environnement (a series of multi-party debates organized in France beginning in 2007 that came to focus on the environment; the encounter in question was devoted to forests): "Produce more wood while

better protecting biodiversity." Besides the oxymoronic absurdity of the slogan, which needs no comment, the fundamental misunderstanding is worth noting: *no one* has ever produced wood – no human, in any case. We capture the products of the dynamics of life that we have at best inflected, or diverted, but we do not produce wood: it is the evolution of gymnosperms and angiosperms, the storehouses of vital solutions incorporated into ligneous organisms enabling them to devour sunshine and make flesh, that produce wood. At the very most, we gather wood.

27 On this point, see Michael D. Wise, *Producing Predators: Wolves, Work and Conquest in the Northern Rockies* (Lincoln: Nebraska University Press, 2016).

28 The historical and philosophical analysis of this ontological event is the object of my current research project, titled "Metaphysics of Production."

29 I can cite the Jivaros described by Descola in their relation to manioc plants (*Beyond Nature and Culture*, p. 343). Let me also cite the words of Big Thunder, an Amerindian of the Wabanakis Nation, referring to the earth as "our mother": "She nourishes us; that which we put into the ground she returns to us, and healing plants she gives us likewise" (T. C. McLuhan, *Touch the Earth: A Self-Portrait of Indian Existence* [New York: Outerbridge & Dienstfrey, 1971], p. 22).

30 Wise, *Producing Predators*.

31 On this point, see Roberte Hamayon, *La chasse à l'âme: Esquisse d'une ethnologie du chamanisme sibérien* (Paris: Société d'ethnologie, 1990), especially her analysis of a Siberian cosmology based on the circulation of flesh among species.

32 See Blanche Magarinos-Rey, *Semences hors-la-loi: La biodiversité confisquée* (Paris: Alternatives, coll. Manifesto, 2015).

33 To see this, it suffices to compare wild wheat grain with the most highly perfected genetically modified wheat: the former is like a mysterious

motor accidentally left behind by extraterrestrials. We can improve it a little, but no one can reproduce it *ex nihilo*: the genetics of photosynthesis and the storing of starch are timeless offerings that are not reproducible.

34 In a sense, it is the ontological and political understanding of the commons as a giving environment, and of humans as living interdependently among the living beings in a giving biotic community, that makes it possible to resist giving in to the "tragedy of the commons" fantasized by Garrett Hardin ("The Tragedy of the Commons," *Science* 162, no. 3859 [1968]: 1243–8). This understanding constitutes a criterion that Elinor Ostrom missed, as I see it, in making the practical management of a commons operational. See Elinor Ostrom, *Governing the Commons: The Evolution of Institutions for Collective Action* (Cambridge University Press, 1990).

35 And the first time that a nonhuman animal is conceivable as a product (making a prehistoric god into a thing, thus making oneself a god – of a different type).

36 Erecting "Man" as a producer would be a dimension of humanism (one element in its foundational act). Man is no longer simply created by God, like the rest of Creation; he acquires a share in divine agency by becoming a producer, while God, in the background, silently leaves the stage. And the environment is not the gift of a donor God but becomes defective matter awaiting human production.

37 How can we believe that any natural beings possess an attribute (the ability to act) that allows them to bring a-nature into being starting from nature? Historically, it has often sufficed to change the postulate according to which humans are natural beings: endow them with a divine essence, and all of a sudden the enigma disappears on its own. It is because they have a spark of divinity that humans, like their Creator, can produce by their actions: create, bring into being another realm.

38 What can be said about the conception of action implicit in this metaphysics? It is "hylomorphic": that is, matter is considered to be pure passivity, pure plasticity, waiting to receive a form that is the product of the human mind and human intention – a form that will be imposed on "natural" matter. But there are other ways to conceive of action. In a Taoist context, for instance, acting consists in welcoming into the self, and, at best, inflecting with infinite delicacy (that of non-acting), the flows of determinations that we receive from outside and that traverse us. By contrast, the concept of freedom in the classical philosophy of Descartes or Kant, for example, holds that action is pure act, an uncaused cause: an absolute starting point in which a subject produces acts by giving form to passive and plastic matter. Kantian freedom is one variant, among others, of the metaphysics of production. In the same way that hylomorphism undervalues the agency of material clay and molds, while overvaluing human acts and the human initiative of giving form, the heirs to this metaphysics conceive of the free act as production while devaluing the flow of conditions received and overvaluing the initiative of the free agent. A free act here is to the flow of external and internal causes that push us to act what a brick is to passive, plastic clay, and what a domesticated sheep is to its ancestor, the wild mouflon.

39 Each of these dimensions has a different history that would warrant study by better historians; they are not asynchronous, but they constitute a disparate configuration whose elements are crystallized in the metaphysics of production.

40 This dimension is hard to trace; fragments of its origin are found in the specific concept of creative action in the Creation story of Judeo-Christian monotheism, while other fragments come from the myth of the demigod in Plato's *Timaeus*, and still others from the Cartesian and Kantian concepts of freedom.

41 Let me note in passing that, if the massification of the practices of vegetable gardening has a metaphysical and civilizational effect, it is this: all practitioners experience *in vivo* the fact that they do not produce their own potatoes. Even apartment-dwellers can experience the absurdity of the metaphysics of production.

42 On pluralism and the common denominator of the agroecologies, see the synthesis by Thierry Doré and Stéphane Ballon, *Les mondes de l'agroécologie* (Versailles: Quae, 2019). On an approach to human technologies that distinguishes between "making something work" and "making do with," see Catherine and Raphaël Larrère, *Penser et agir avec la nature* (Paris: La Découverte, 2015).

43 See, especially, Geerat Vermeij, *Evolution and Escalation: An Ecological History of Life* (Princeton University Press, 1987).

44 See Baptiste Morizot, "Ce que le vivant fait au politique," in Frédérique Aït-Touati and Emanuele Coccia, eds., *Le cri de Gaïa: Penser la Terre avec Bruno Latour* (Paris: La Découverte, 2021), pp. 77–118.

45 We must keep in mind an important point (I owe this idea to Sébastien Blache): all agriculture is based on an environmental disequilibrium that it has created and of which it is the heir. To bring together on a small surface area a large density of apple trees that produce appetizing fruit in abundance is to create an environmental imbalance, a situation that does not occur in spontaneous ecosystems. This practice inevitably attracts predators, and favors proliferations, just as the fact of bringing sheep that have been rendered docile together on a plain attracts wolves. What is behind attacks by aphids or slugs on a vegetable garden, the invasion of "pests," is always the environmental imbalance created by agriculture. The origin of such attacks is invisible, but it haunts the agricultural space year after year, century after century. This is not a moral observation, but rather an instance of historical awareness. The whole question then becomes

how to react to this imbalance. Should one respond to it by aggressive action in return that accentuates it, increases it, and draws one's agricultural practice into a spiral of mutual aggression, or should one compensate for it, smooth it over, minimize it, and favor the dynamics of life?

46 Facundo Ibañez, Woo Young Bang, Leonardo Lombardini, and Luis Cisneros-Zevallos, "Solving the Controversy of Healthier Organic Fruit: Leaf Wounding Triggers Distant Gene Expression Response of Polyphenol Biosynthesis in Strawberry Fruit (*Fragaria x ananassa*)," *Scientific Reports*, no. 9 (2019), http://doi.org/10.1038/s41598-019-55033-w.

47 A study carried out in 2014 at the University of Newcastle had already shown, through meta-analysis, that organic fruits and vegetables have higher concentrations (from 18 percent to 69 percent) of antioxidants (such as phenolic acids, flavanones, stilbenes, flavones, flavenols, and anthocyanins): Marcin Barański, Dominika Srednicka-Tober, Nikolaos Volakakis, et al., "Higher Antioxidant and Lower Cadmium Concentrations and Lower Incidence of Pesticide Residues in Organically Grown Crops: A Systematic Literature Review and Meta-Analyses," *British Journal of Nutrition* 112, no. 5 (September 9, 2014): 794–811, www.ncbi.nlm.nih.gov/pmc/articles/PMC4141693.

48 Ibañez et al., "Solving the Controversy."

49 Ibañez et al., "Solving the Controversy."

50 On this point, see Charlotte Brives, "Politiques de l'amphibiose: La guerre contre les virus n'aura pas lieu," *Le Média* (March 21, 2020): https://lemediatv.fr/articles/2020politiques-de-lamphibiose-la-guerre-contre-les-virus-naura-pas-lieu-ACcrS8oIQsOuLmmvfx2aQ.

51 See Claude Combes, *Parasitism: The Ecology and Evolution of Intimate Interactions*, trans. Isaure de Buron and Vincent A. Connors (University of Chicago Press, [2001] 2018).

52 On this point, see Baptiste Morizot, "Adjusted Consideration," the epilogue to *Ways of Being Alive*, trans. Andrew Brown (Cambridge: Polity, 2022).
53 At its headquarters in Gotheron, the INRA houses an experimental unit for integrated research in agronomy.
54 On this point, see, for example, Christophe Bonneuil, Gilles Denis, and Jean-Luc Mayaud, eds., *Sciences, chercheurs et agriculture: Pour une histoire de la recherche agronomique* (Paris: L'Harmattan, 2008).
55 Nineteenth-century scientist Justus von Liebig articulated the "Law of the Minimum," which holds that, even if most needed resources are plentiful, a plant's growth will not improve unless the amount of the scarcest resource is increased.
56 See Aurélien Gabriel Cohen, "Des lois agronomiques à l'enquête agroécologique: Esquisse d'une épistémologie de la variation dans les agrosystèmes," *Tracés* 17, no. 33 (2017): 51–72.
57 Michel Jay is a former researcher at the Centre technique interprofessionnel des fruits et légumes (Interprofessional Technology Center for Fruits and Vegetables, CTIFL).
58 Here we find the fundamental thesis of the incommensurability of the different values involved with the single metric of money, which underlies ecologic economics. On this point, see Joan Martinez Alier, *The Environmentalism of the Poor: A Study of Ecological Conflicts and Valuation* (Cheltenham, UK: Edward Elgar Publishing, 2002).
59 "As we felled and burned the forests, so we burned, plowed, and overgrazed the prairies. We came with visions, but not with sight. We did not see or understand where we were or what was there, but destroyed what was there for the sake of what we desired. And the desire was always native to the place we left behind" (Wendell Berry, "The Native Grasses and What They Mean," in Wendell Barry, *The Gift*

*of Good Land: Further Essays Cultural and Agricultural* [San Francisco: North Point Press, 1981], p. 82).

60 In this connection, see Marc Dufumier and Olivier Le Naire, *L'agroécologie peut nous sauver* (Arles: Actes Sud, 2019).

61 For certain milieus, free evolution is hard to justify: these milieus would lose in functionality and in biodiversity if we were to withdraw all active management. Highlands, wooded countrysides, and orchards are milieus that have been co-produced by human activities. These activities, unlike the industrial exploitation of forests, are no longer anthropic forcings that prevent a milieu from thriving (as with the forests in France), they are elements of the very identity of the area: when forcings are your identity, they are no longer forcings. In these cases, free evolution makes no sense. Nevertheless, in these milieus, a management style in the spirit of rekindling the embers can be defended, a style that is based on trust in the dynamics of life. This type of management will valorize the "helping hand" that restores autonomy to a milieu, as opposed to an arrangement that is always subject to entropy and that requires constant reactualization through human intervention. Everything is directed by trust and the helping-hand type of action, with interventions differing in degree depending on the milieu.

62 See the "Paysans de nature" (Naturalist Farmers) network started by farmers and the League for the Protection of Birds in the Vendée region (LPO Vendée): www.paysansdenature.fr.

63 The propositions set forth here are not just idle speculations; they are maps designed to orient us on the ground, maps that can be activated in order to forge concrete political alliances among different uses of land. I recently had an opportunity to test them among members of the Farmers' Confederation and the board of directors of ASPAS, at a meeting organized by the two associations to discuss their conflict. The outcome was not obvious: it was a question of reestablishing dialogue

following the publication of a motion passed by the Drôme Farmers' Confederation against the Réserves de Vie Sauvage. It was unclear to what extent it was appropriate or desirable for the various actors around the table to speak of an alliance. I was struck by the fact that, as soon as an actor in the discussion fell back on dualisms (exploitation is opposed to sanctuarization; all human activity degrades the milieu; the destiny of the earth is to be cultivated by traditional farmers), the conflicts became crystallized to the point of making dialogue impossible. In contrast, as soon as we managed to bring to light the common enemy (namely, extractivist and industrial agriculture and the artificialization of soils) and the relation to the living world shared by naturalists and farmers (trust in the dynamics of life), the space for discussion opened back up. In this way, it became possible to formulate a sort of double dependency: the traditional and agroecological farming often defended by the Farmers' Confederation was the only form of agriculture compatible with the defense of wild milieus supported by ASPAS; the traditional agriculture defended by the Confederation does not "produce" but, rather, welcomes and inflects the wild dynamics (farmers fighting "modern" agriculture grasp this immediately); finally, these dynamics are reinvigorated by the existence of hearths of free evolution woven together with sustainable agricultural activities. The Confederation and ASPAS also came together around a common way of defending the living world: in both instances (reserves without exploitation and traditional agriculture), the actors agreed on the fact that it was a matter not of protecting nature but of defending multispecies milieus of life. At the end of the meeting, the presumed opposition in the practices of the two groups (forbidding exploitation on one side, practicing exploitation on the other) had given way to an at least tacit recognition of two positions that were nevertheless on the *same* front – that of sustainable relations with the dynamics that exceed us and produce us.

As for the political results that can be expected from the meeting, quite frankly, only time will tell.

64 One thing I am exploring here is whether it is possible to think through the question of alliances with the living world by using Ernesto Laclau's concepts concerning the construction of hegemonies through the aggregation of struggles and stakes. It is not necessarily a matter of constructing a universal principle, but rather, first of all, a richer political subject – a larger, composite political body. It is a matter of envisioning sustainable agricultures (agroecologies, organic agriculture, permaculture) aggregated into a powerful entity that encompasses pollinator bees, pregnant women, children at risk of higher mortality rates owing to the lower vitamin intake induced by the disappearance of pollinators, beekeepers whose practices are sustainable – allies and aggregates *against* other uses and interest groups. This aggregation does not take place following the model of atomized individuals calculating their own interests; instead, it follows the hybrid model of Spinoza's individual, who harmonizes shared powers within the ecological community, and the model of hegemonic aggregation in political speech and struggles. We could thus imagine the operational logic of a political ecology based on objective alliances: these alliances would need to be spurred, aggregated, and expanded by actors monitoring constitutive relations, observing their secondary effects on the entire network, and anticipating the risks. It would be a matter of a multispecies hegemonic aggregation that creates polycephalic political bodies that are powerful in new ways. On this point, see Federico Tarragoni, "Vers une logique générale du politique: Identités, subjectivations et émancipations chez Laclau," *Revue du MAUSS permanente*, January 25, 2015: wwwjournaldumauss.net/spip.php?page=imprimer&id_article=1203. See also Ernesto Laclau, *On Populist Reason* (London: Verso, 2008).

## 5 Making Maps Differently: Dealing with Disagreements

1 Edward O. Wilson, *Half-Earth: Our Planet's Fight for Life* (New York: Liveright, 2016).
2 See Isaac Kantor, "Ethnic Cleansing and America's Creation of National Parks," *Public Land and Resources Law Review* 28 (2007): 41–64.
3 In this connection, see Malcom Ferdinand, *Decolonial Ecology: Thinking from the Caribbean World*, trans. Anthony Paul Smith (Hoboken, NJ: Blackwell, forthcoming 2022).
4 This problem is analogous to the one posed by mechanisms of compensation: if we finance the replanting of trees on a few faraway acres, we can continue to destroy locally. Sanctuarization ultimately serves exploitation, legitimizing it and serving to salve the conscience of its practitioners. The dualism implied by sanctuarization is a constitutive arrangement of greenwashing.
5 In reality, most have already understood this, and they have been showing the way for years now: for examples in France, I can point to the Férus association with Pastoraloup, and their simultaneous defense of a wild animal – the wolf – and of sustainable pastoralism. Another example would be the League for the Protection of Birds and its involvement with the projects of Naturalist Farmers.
6 This is a way of restoring meaning at the heart of national parks: they can be interpreted as the first zones of free evolution (when they have not been subjected to too many compromises) and defended in their existence (if not in their discourse) by the arguments proposed here. This is why the struggle to maintain these parks and the employees who manage them is indispensable at present.
7 One of the challenges of a more "diplomatic" forest management style is to initiate a "withdrawal of anthropic forcings," according to Marina Fischer-Kowalski: see Marina Fischer-Kowalski and Thomas

Macho, *Gesellschaftlicher Stoffwechsel und Kolonisierung von Natur: Ein Versuch in Sozialer Ökologie* (Amsterdam: G+B Verlag Fakultas, 1997). In this connection, Guillaume Christen has written: "Following [Fischer-Kowalski's] perspective, Frédéric Goulet and Dominique Vinck define practices relating to nature characterized by management 'through withdrawal.' These gentler modes of management have been tried most notably in natural silviculture, the objectives of which consist in removing modes of intervention that act directly on forest ecosystems (plantations, treatments applied to young plants) in order to favor the dynamics and the potentialities of the milieu, that is, to optimize the cycle of repopulation of the essences in place. The forest ecosystem takes on a new status: its processes go back to being useful and functional objects in the logic of forest production. Rather than 'improving' the forest, this model seeks to understand the potentialities of the milieu and to steer their auxiliary role and function in the productive processes" ("La place de la nature spontanée à l'ère de l'anthropocène: Le regard des sciences sociales," *Naturalité, la lettre de Forêts Sauvages*, no. 20 [May 2019]: 18–19). See also Frédéric Goulet and Dominique Vinck, "L'innovation par retrait: Contribution à une sociologie du détachement," *Revue française de sociologie* 53, no. 2 (2012): 195–224.

8 This point is clearly made in *Le temps des forêts*, a documentary film by François-Xavier Drouet released in 2018.

9 It must be distinguished from dynamics through which a milieu that is no longer managed finds spontaneous ecological trajectories once again, like the fate of wastelands that gradually become reforested. This spontaneous phenomenon is starting to be known as feralization (see Annick Schnitzer and Jean-Claude Génot, *La nature férale ou le retour du sauvage* [Geneva: Jouvence, 2020]).

10 Michael Soulé and Reed Noss, "Rewilding and Biodiversity: Complementary Goals for Continental Conservation," *Wild Earth* (Fall 1998): 19–28.
11 In this connection, see the figures of the group Pour une autre PAC concerning the effects of artificializing both rural lands and "wild" milieus: www.pouruneautrepac.eu/comprendre-la-pac/reformer-la-pac.
12 Terre de Liens is an association that seeks to help farmers acquire land and to preserve agricultural terrain – in particular, by fighting against speculation in real estate and against the artificialization of soils.
13 This consists in a certain way of combining what Anglo-Saxon conservation biology calls *land sparing* (putting certain spaces into "protection") and *land sharing* (coexistence with the biodiversity present outside of the protected spaces).

## 6 Conclusion: The Living World Defends Itself

1 On this point, see Georges Canguilhem, "The Living and Its Milieu," in *Knowledge of Life*, trans. Paola Marrati and Todd Meyers (New York: Fordham University Press, [1952] 2008), pp. 98–120. Isabelle Stengers formulates an analogous conclusion in a luminous way, starting from her reading of Alfred North Whitehead, in *Civiliser la modernité? Whitehead et les ruminations du sens commun* (Dijon: Les Presses du réel, 2017), pp. 140–2.
2 In his important book *Infravies: Le vivant sans frontières* (Paris: Seuil, 2019), Thomas Heams calls into question the abrupt break between organic and inorganic forms. He does this by focusing on a major intermediary vestibule: the infraliving. From an ontological viewpoint, he is right; there is indeed a liminal continent of "infralives" between the living cellular being and inorganic matter – but the distinctions must not be made absolute; they are tools in relation to problems. If life must

not be reduced to cellular life but opened up to this infraliving world, it would be philosophically erroneous to wager on these discoveries and announce that there is no difference between organic and inorganic entities, that "everything is alive." The existence of infraliving entities does not refute the existence of nonliving entities; inorganic matter does exist: it takes up the vast majority of space in the known cosmos; it is not alive. In contrast, there exists a rich, ample, and underestimated grey zone of infralives, which are not cellular in themselves but which are on the borderline of cellular life and make cellular life forms possible. And it remains very difficult, even impossible, when we go down to the infracellular level, to decree where life begins. But that matters little here: when one is part of life, one is concerned with one's own existence.

3 On this point, see Daniel Chamovitz's essential book, *What a Plant Knows: A Field Guide to the Senses* (New York: Scientific American / Farrar, Straus and Giroux, 2017).

4 Today, the culpability of the Moderns for what we have done to our world is such that it is activating on a massive scale a political drive that was invented to respond to injustices between humans: a drive to be inclusive and egalitarian. From the perspective of that drive, people may be shocked by my arguments, seeing in them a sort of discrimination against the "poor" stones, the clouds, the ozone layer. I am not saying that these are without value; it is not a matter, here, of distributing badges of axiological consistency, but of delineating the camp of the allies, the minimal political unit. There are a thousand ways to politicize clouds, rocks, and technological objects, of granting them importance, and these ways are necessary – but not for solving the specific problem I am raising here. On this strange egalitarianism, see, for example, Jane Bennett, *Vibrant Matter: A Political Ecology of Things* (Durham, NC: Duke University Press, 2009).

5 Claude Lévi-Strauss, *The Elementary Structures of Kinship*, ed. Rodney Needham, rev. edn., trans. James Harle Bell and John Richard von Sturmer (Boston: Beacon Press, [1949/1967] 1969), "Preface to the Second Edition," p. xxix.

6 On this point, see Stéphane Madelrieux, *La philosophie de John Dewey* (Paris: Vrin, 2016).

7 See Virginie Maris, *La part sauvage du monde* (Paris: Seuil, 2018).

8 For a catalogue of all that life does to make the world inhabitable, see Timothy M. Lenton and Sébastien Dutreuil, "What Exactly Is the Role of Gaia?" in Bruno Latour and Peter Weibel, eds., *Critical Zones: The Science and Politics of Landing on Earth* (Cambridge, MA: MIT Press, 2020), pp. 168–75.

9 Certain micro-organisms, the glaciogenic bacteria, have the ability to support the formation of ice crystals in the atmosphere; these crystals concentrate water drops and thus form clouds. On this topic, see Tina Šanti-Temkiv, Kai Finster, Thorsten Dittmar, et al., "Hailstones: A Window into the Microbial and Chemical Inventory of a Storm Cloud," *PLOS One* 8, no. 1 (2013): e53550, https://doi.org/10.1371/journal.pone.0053550.

10 Why, then, do we not feel the gratitude logically owed to these bacteria for every mouthful of water we swallow? We have inherited a metaphysical schema that defuses every possibility of gratitude when the gift is not intentional; this is part of the monotheist legacy. Every gift that is not intended is not considered to be a gift but rather a given, almost our due – something material that is present only for us to take. Making gratitude possible for the non-intentional gifts that keep us alive unblocks one decisive barrier in our relation to the living world.

11 This picture must, nevertheless, not be made to appear overly peaceful: living beings make the world inhabitable for the existing communities

of living beings, but not in consensual unanimity. Certain living beings are disappearing owing to the effects produced by other living beings; the evolutionary trajectories of communities always imply that certain populations and species are fading away to the benefit of others. There is no great political community preserved by the living world in general, but only trajectories, and a sort of minimal condition that is inhabitability for life in general, such as it gives itself to us at this moment in its evolutionary history. At no point has life on earth desired or intended to make the world inhabitable – that is simply something it does to the world. And it is inhabitable for us because we are part of it, we have coevolved as one thread in its weaving.

12 Gilles Deleuze and Claire Parnet, *Dialogues II*, rev. edn., trans. Hugh Tomlinson and Barbara Habberjam (New York: Columbia University Press, [1989] 1996), p. 45.

13 James E. Lovelock, "The Earth Is Not Fragile," in Bryan Cartledge, ed., *Monitoring the Environment: The Linacre Lectures 1990–91* (Oxford University Press, 1992), pp. 120–1.

14 Lovelock, "The Earth Is Not Fragile," p. 121.

15 Lévi-Strauss hypothesized that, "from the first [*Homo sapiens*] alone had claimed to personify culture as opposed to nature, and to remain now, except for those cases where it can totally bend it to its will, the sole embodiment of life as opposed to inanimate matter. By this hypothesis, the contrast of nature and culture would be neither a primeval fact, nor a concrete aspect of universal order. Rather it should be seen as an artificial creation of culture, a protective rampart thrown up around it because it only felt able to assert its existence and uniqueness by destroying all the lengths that lead back to its original association with the other manifestations of life" (Lévi-Strauss, *Elementary Structures of Kinship*, p. xxix).

# Works Cited

Alier, Joan Martinez. *The Environmentalism of the Poor: A Study of Ecological Conflicts and Valuation*. Cheltenham, UK: Edward Elgar Publishing, 2002.

Athanaze, Pierre. "Des réserves de vie sauvage." *Naturalité, la lettre de Forêts Sauvages*, no. 13 (February 2014): 2.

Barański, Marcin, Dominika Srednicka-Tober, Nikolaos Volakakis, et al. "Higher Antioxidant and Lower Cadmium Concentrations and Lower Incidence of Pesticide Residues in Organically Grown Crops: A Systematic Literature Review and Meta-Analyses." *British Journal of Nutrition* 112, no. 5 (September 9, 2014): 794–811. www.ncbi.nlm.nih.gov/pmc/articles/PMC4141693.

Barnosky, Anthony D. "Megafauna Biomass Tradeoff as a Driver of Quaternary and Future Extinctions." *PNAS* 12, no. 5, suppl. 1 (2008): 11543–8.

Bennett, Jane. *Vibrant Matter: A Political Ecology of Things*. Durham, NC: Duke University Press, 2009.

Berry, Wendell. "The Native Grasses and What They Mean." In Wendell Berry, *The Gift of Good Land: Further Essays Cultural and Agricultural*, pp. 77–97. San Francisco: North Point Press, 1981.

Bird-David, Nurit. "The Giving Environment." *Current Anthropology* 31, no. 2 (April 1990): 189–96.

Bonneuil, Christophe, Gilles Denis, and Jean-Luc Mayaud, eds., *Sciences, chercheurs et agriculture: Pour une histoire de la recherche agronomique*. Paris: L'Harmattan, 2008.

## Works Cited

Brives, Charlotte. "Politiques de l'amphibiose: La guerre contre les virus n'aura pas lieu." *Le Média* (March 21, 2020). https://lemediatv.fr/articles/2020politiques-de-lamphibiose-la-guerre-contre-les-virus-naura-pas-lieu-ACcrS8oIQsOuLmmvfx2aQ.

Callicott, J. Baird. *Earth's Insights: A Multicultural Survey of Ecological Ethics from the Mediterranean Basin to the Australian Outback*. Berkeley: University of California Press, 1994.

Canguilhem, Georges. "The Living and Its Milieu." In *Knowledge of Life*, trans. Paola Marrati and Todd Meyers, pp. 98–120. New York: Fordham University Press, [1952] 2008.

Canguilhem, Georges. "The Normal and the Pathological." In *Knowledge of Life*, trans. Paola Marrati and Todd Meyers, pp. 121–33. New York: Fordham University Press, [1952] 2008.

Cauvin, Jacques. *The Birth of the Gods and the Origins of Agriculture*, trans. Trevor Watkins. Cambridge University Press, [1997] 2000.

Chamovitz, Daniel. *What a Plant Knows: A Field Guide to the Senses*. New York: Scientific American / Farrar, Straus and Giroux, 2017.

Charbonnier, Pierre. *Abondance et liberté: Une histoire environnementale des idées politiques*. Paris: La Découverte, 2020.

"Charte des Réserves de Vie Sauvage." https://aspas-reserves-vie-sauvage.org.

Childe, Vere Gordon. *Man Makes Himself*. New York: Oxford University Press, 1939.

Christen, Guillaume. "La place de la nature spontanée à l'ère de l'anthropocène: Le regard des sciences sociales." *Naturalité, la lettre de Forêts Sauvages*, no. 20 (May 2019): 16–19.

Cochet, Gilbert, and Stéphane Durand. *Ré-ensauvageons la France: Plaidoyer pour une nature sauvage et libre*. Arles: Actes Sud, 2018.

## Works Cited

Cohen, Aurélien Gabriel. "Des lois agronomiques à l'enquête agroécologique: Esquisse d'une épistémologie de la variation dans les agrosystèmes." *Tracés* 17, no. 33 (2017): 51–72.

Combes, Claude. *Parasitism: The Ecology and Evolution of Intimate Interactions*, trans. Isaure de Buron and Vincent A. Connors. University of Chicago Press, [2001] 2018.

Darwin, Charles. *On the Various Contrivances by Which British and Foreign Orchids Are Fertilized by Insects, and on the Good Effects of Intercrossing*. London: J. Murray, 1862.

Darwin, Charles. *The Origin of Species*. New York: P. F. Collier & Son, [1859] 1909.

Deleuze, Gilles, and Claire Parnet. *Dialogues II*, rev. edn., trans. Hugh Tomlinson and Barbara Habberjam. New York: Columbia University Press, [1989] 1996.

Descartes, René. *Meditations on First Philosophy*. In René Descartes, *The Philosophical Works of Descartes*, trans. Elizabeth S. Haldane. Cambridge University Press, [1641] 1911.

Descola, Philippe. *Beyond Nature and Culture*, trans. Janet Lloyd. University of Chicago Press, 2013.

Doré, Thierry, and Stéphane Ballon. *Les mondes de l'agroécologie*. Versailles: Quae, 2019.

Drayton, Richard. *Nature's Government*. New Haven, CT: Yale University Press, 2000.

Duarte, Carlos M., Susana Agusti, Edward Barbier, et al. "Rebuilding Marine Life." *Nature*, no. 580 (2020): 39–51.

Dufumier, Marc, and Olivier Le Naire. *L'agroécologie peut nous sauver*. Arles: Actes Sud, 2019.

Dutreuil, Sébastien. "Quelle est la nature de la Terre?" In Frédérique Aït-Touati and Emanuele Coccia, eds., *Le cri de Gaïa: Penser la Terre avec Bruno Latour*, pp. 17–65. Paris: La Découverte, 2021.

## Works Cited

Erb, Karl Heinz, T. Kastner, C. Plutzar, et al. "Unexpectedly Large Impact of Forest Management and Grazing on Global Vegetation Biomass." *Nature*, no. 553 (2018): 73–6.

Falbet, Philippe. Open letter to Confédération Paysanne de l'Ariège. Aspet (Haute-Garonne), May 17, 2019.

Ferdinand, Malcom. *Decolonial Ecology: Thinking from the Caribbean World*, trans. Anthony Paul Smith. Cambridge: Polity, 2022.

Fischer-Kowalski, Marina, and Thomas Macho. *Gesellschaftlicher Stoffwechsel und Kolonisierung von Natur: Ein Versuch in Sozialer Ökologie*. Amsterdam: G+B Verlag Fakultas, 1997.

Foer, Jonathan Safran. *We Are the Weather: Saving the Planet Begins at Breakfast*. New York: Farrar, Straus and Giroux, 2019.

Génot, Jean-Claude. *La nature malade de la gestion*. Paris: Le Sang de la terre, 2008.

Godet, Laurent. "La survie du monde vivant doit passer avant le développement économique." Interview. *Usbek et Rica*, September 10, 2018. https://usbeketrica.com/fr/article/la-survie-du-monde-vivant-doit-passer-avant-le-developpement-economique.

Godet, Laurent, and Vincent Devictor. "What Conservation Does." *Trends in Ecology & Evolution* 33, no. 10 (October 2018): 720–30.

Goulet, Frédéric, and Dominique Vinck. "L'innovation par retrait: Contribution à une sociologie du détachement." *Revue française de sociologie* 53, no. 2 (2012): 195–224.

Gruzinski, Serge. *The Mestizo Mind: The Intellectual Dynamics of Colonization and Globalization*, trans. Deke Dunsinberre. New York: Routledge, 2002.

Hache, Émilie. *Ce à quoi nous tenons*. Paris: La Découverte, 2014.

Hache, Émilie. "Le vol du sang: Relire la théologie chrétienne à l'aune de Gaïa." February 6, 2020. www.youtube.com/watch?v=9f-OyHRMMpw&feature=youtu.be.

*Works Cited*

Hamayon, Roberte. *La chasse à l'âme: Esquisse d'une ethnologie du chamanisme sibérien*. Paris: Société d'ethnologie, 1990.

Hardin, Garrett. "The Tragedy of the Commons." *Science* 162, no. 3859 (1968): 1243–8.

Heams, Thomas. *Infravies: Le vivant sans frontières*. Paris: Seuil, 2019.

Ibañez, Facundo, Woo Young Bang, Leonardo Lombardini, and Luis Cisneros-Zevallos. "Solving the Controversy of Healthier Organic Fruit: Leaf Wounding Triggers Distant Gene Expression Response of Polyphenol Biosynthesis in Strawberry Fruit (*Fragaria x ananassa*)." *Scientific Reports*, no. 9 (2019). http://doi.org/10.1038/s41598-019-55033-w.

Kantor, Isaac. "Ethnic Cleansing and America's Creation of National Parks." *Public Land and Resources Law Review* 28 (2007): 41–64.

Laclau, Ernesto. *On Populist Reason*. London: Verso, 2008.

Larrère, Catherine, and Raphaël Larrère. *Penser et agir avec la nature*. Paris: La Découverte, 2015.

Larrère, Raphaël. "Le réparateur, l'ingénieur ou le thérapeute?" *Sciences, eaux & territoires* 3, no. 24 (2017): 16–19.

Laussel, Pascale, Marjolaine Boitard, and Gaëtan du Bus de Warnaffe. *Agir ensemble en forêt: Guide pratique, juridique et humain*. Paris: Charles-Léopold Mayer, 2018.

Le Goff, Jacques. "Ecclesiastical Culture and Folklore in the Middle Ages: Saint Marcellus of Paris and the Dragon." In *Time, Work & Culture in the Middle Ages*, trans. Arthur Goldhammer, pp. 159–88. University of Chicago Press, 1980.

Le Goff, Jacques, with Emmanuel Le Roy Ladurie. "Mélusine maternelle et défricheuse." *Annales. Économies, sociétés, civilisations* 25, nos. 3–4 (1971): 587–622.

Lenton, Timothy M., and Sébastien Dutreuil, "What Exactly Is the Role of Gaia?" In Bruno Latour and Peter Weibel, eds., *Critical Zones: The*

*Science and Politics of Landing on Earth*, pp. 168–75. Cambridge, MA: MIT Press, 2020.

Lévi-Strauss, Claude. *The Elementary Structures of Kinship*, ed. Rodney Needham, rev. edn., trans. James Harle Bell and John Richard von Sturmer. Boston: Beacon Press, [1949/1967] 1969.

Lovelock, James. "The Earth Is Not Fragile." In Bryan Cartledge, ed., *Monitoring the Environment: The Linacre Lectures 1990–91*, pp. 105–22. Oxford University Press, 1992.

Lovelock, James, and Lynn Margulis, "Biological Modulation of the Earth's Atmosphere." *Icarus* 21, no. 4 (1974): 471–89.

Madelrieux, Stéphane. *La philosophie de John Dewey*. Paris: Vrin, 2016.

Magarinos-Rey, Blanche. *Semences hors-la-loi: La biodiversité confisquée*. Paris: Alternatives, coll. Manifesto, 2015.

Maris, Virginie. *La part sauvage du monde*. Paris: Seuil, 2018.

Marris, Emma. *Rambunctious Garden: Saving Nature in a Post-Wild World*. New York: Bloomsbury, 2013.

Maurel, Lionel. "Communs & non-humains (1ère partie): Oublier les 'ressources' pour ancrer les Communs dans une 'communauté biotique.'" Blog S. I. Lex, January 10, 2019. https://scinfolex.com/2019/01/10/communs-non-humains-1ere-partie-oublier-les-ressources-pour-ancrer-les-communs-dans-une-communaute-biotique.

Maurel, Lionel. "La propriété privée au secours des forêts ou les paradoxes des nouveaux communs sylvestres." Blog S. I. Lex, August 19, 2019. https://scinfolex.com/2019/08/19/la-propriete-privee-au-secours-des-forets-ou-les-paradoxes-des-nouveaux-communs-sylvestres.

McLuhan, T. C. *Touch the Earth: A Self-Portrait of Indian Existence*. New York: Outerbridge & Dienstfrey, 1971.

Merchant, Carolyn. *The Death of Nature: Women, Ecology and the Scientific Revolution*. San Francisco: HarperOne, 1990.

## Works Cited

Morizot, Baptiste. "Ce que le vivant fait au politique: La spécificité des vivants en contexte de métamorphoses environnementales." In Frédérique Aït-Touati and Emanuele Coccia, eds., *Le cri de Gaïa: Penser la Terre avec Bruno Latour*, pp. 77–118. Paris: La Découverte, 2021.

Morizot, Baptiste. *Ways of Being Alive*, trans. Andrew Brown. Cambridge: Polity, 2022.

*Naturalité, la lettre de Forêts Sauvages.* www.forets-sauvages.fr/web/foret sauvages/100-naturalite-la-lettre-de-forets-sauvages.php.

Nietzsche, Friedrich. *The Twilight of the Idols.* Trans. Duncan Large. New York: Oxford University Press, 1998.

Ostrom, Elinor. *Governing the Commons: The Evolution of Institutions for Collective Action.* Cambridge University Press, 1990.

Patel, Raj, and Jason Moore. *A History of the World in Seven Cheap Things: A Guide to Capitalism, Nature, and the Future of the Planet.* Oakland: University of California Press, 2017.

Perlès, Catherine. "Pourquoi le Néolithique? Analyse des théories, évolution des perspectives." In Jean-Pierre Poulain, ed., *L'homme, le mangeur, l'animal, qui nourrit l'autre?* pp. 16–29. Paris: Observatoire Cidil des habitudes alimentaires, coll. Cahiers de l'OCHA, no. 12, 2007.

Persuy, Alain. "Pour des forêts vivantes, résistance!!!" *Naturalité, la lettre de Forêts Sauvages*, no. 18 (December 2018): 8.

Regnery, Baptiste, Denis Couvet, Loren Kubarek, Jean-François Julien, and Christian Kerbiriou. "Tree Microhabitats as Indicators of Bird and Bat Communities in Mediterranean Forests." *Ecological Indicators* 34, no. 1 (November 2013): 221–30.

Regnery, Baptiste, Yoan Paillet, Denis Couvet, and Christian Kerbiriou. "Which Factors Influence the Occurrence and Density of Tree Microhabitats in Mediterranean Oak Forests?" *Forest Ecology and Management* 295 (May 1, 2013): 118–25.

# Works Cited

Robert, Alexandre, Colin Fontaine, Simon Veron, et al. "Fixism and Conservation Science." *Conservation Biology* 31, no. 4 (August 2017): 781–8. https://pubmed.ncbi.nlm.nih.gov/27943401.

Rosny, J.-H. *The Quest for Fire: A Novel of Prehistoric Times*, translated from the French. New York: Pantheon Books, [1911] 1967.

Šanti-Temkiv, Tina, Kai Finster, Thorsten Dittmar, et al. "Hailstones: A Window into the Microbial and Chemical Inventory of a Storm Cloud." *PLOS One* 8, no. 1 (2013): e53550. https://doi.org/10.1371/journal.pone.0053550.

Sarrazin, François, and Jane Lecomte. "Evolution in the Anthropocene." *Science* 351, no. 6276 (February 26, 2016): 922–3.

Schnitzer, Annick, and Jean-Claude Génot. *La nature férale ou le retour du sauvage*. Geneva: Jouvence, 2020.

Smil, Vaclav. "Harvesting the Biosphere: The Human Impact." *Population and Development Review* 37, no. 4 (2011): 613–36.

Soulé, Michael, and Reed Noss. "Rewilding and Biodiversity: Complementary Goals for Continental Conservation." *Wild Earth* (Fall 1998): 19–28.

Stengers, Isabelle. *Civiliser la modernité? Whitehead et les ruminations du sens commun*. Dijon: Les Presses du réel, 2017.

Tarragoni, Federico. "Vers une logique générale du politique: Identités, subjectivations et émancipations chez Laclau." *Revue du MAUSS permanente* (January 25, 2015): wwwjournaldumauss.net/spip.php?page=imprimer&id_article=1203.

Thomas, Chris. *Inheritors of the Earth: How Nature Is Thriving in an Age of Extinction*. London: Penguin, 2017.

Vermeij, Geerat. *Evolution and Escalation: An Ecological History of Life*. Princeton University Press, 1987.

Warde, Paul. "The Idea of Improvement, c. 1520–1700." In Richard W. Hoyle, ed., *Custom, Improvement and the Landscape in Early Modern Britain*, pp. 127–48. Farnham and Burlington: Ashgate, 2011.

## Works Cited

Warnaffe, Gaëtan du Bus de, and Sylvain Angerand. *Gestion forestière et changement climatique: Une nouvelle approche de la stratégie nationale d'atténuation.* January 2020. www.alternativesforestieres.org/IMG/pdf/synthese-web-rapport-foret-climat-fern-canopee-at.pdf.

Wilson, Edward O. *Half-Earth: Our Planet's Fight for Life.* New York: Liveright, 2016.

Wise, Michael D. *Producing Predators: Wolves, Work and Conquest in the Northern Rockies.* Lincoln: Nebraska University Press, 2016.

Wood, Neal. *John Locke and Agrarian Capitalism.* Berkeley: University of California Press, 1984.